BEACHCRAFTS, TOO!

An All-New Companion Guide to Beachcraft Bonanza
for Parents, Teachers, Students, Craftspeople, Children and Beachlovers
YEAR-ROUND ARTS, CRAFTS, PUZZLES, EXPERIMENTS,
AND ACTIVITIES FOR THE SEASHORE

*May these ideas be
of lasting value —
Brian Heinz*

written and illustrated by

Brian J. Heinz

Published by

BALLYHOO BOOKS
P.O. Box 534
Shoreham, NY 11786

BEACHCRAFTS, TOO!

First printing: January, 1988

ISBN: 0-936335-01-7

Library of Congress Cataloging-in-Publication Data

Heinz, Brian, 1946-
    Beachcrafts, too.

    Bibliography: p.
    Summary: Presents craft and art projects, games, and nature study experiments using the various items found on the seashore. Includes beach pebble polishing, sand casting, scallop wind chimes, mapping a tidal creek, and seaside picture bingo.
    1. Nature craft--Juvenile literature. 2. Beaches--Study and teaching--Juvenile literature. 3. Seashore biology--Study and teaching--Juvenile literature.
    [1. Seashore biology. 2. Nature craft. 3. Handicraft]

    I. Title.
    TT160.H384 1988   508.314'6   87-35232
    ISBN 0-936335-01-7 (pbk.)

Printed in the United States of America

For Michael,

Jamie,

Kathleen,

Brian,

Liam,

and Moira

ACKNOWLEDGEMENTS

The text of this book is set in 14 pt. Benguiat Gothic Book.
The display font is Bauhaus Bold.

For information on ordering the companion guide to this book,
"Beachcraft Bonanza," contact:

Ballyhoo Books
P.O. Box 534
Shoreham, NY 11786

CONTENTS

AUTHOR'S NOTE

BEACHCRAFTS, TOO! was born in the wake of the popularity and success of its predecessor, BEACHCRAFT BONANZA, published in 1986.

In BEACHCRAFTS, TOO! you will find an all-new and varied array of stimulating, enjoyable activities designed to instill a sense of wonder, appreciation, and respect for our marine environments. The activities play a dual role, both recreational and instructional, and include arts, crafts, science experiments, puzzles, scavenger hunts, language arts activities, and all-around observational diversions.

People tend to care for, and maintain, those things in their lives which they have deemed important for themselves. Their environment is important.

BEACHCRAFTS, TOO! is a vehicle for the discovery and realization that a marine environment is truly a special place, but a delicate place, that requires us to be considerate, lifelong caretakers of an important and valuable resource.

I thank all the children, parents, and teachers who so eagerly gobbled up BEACHCRAFT BONANZA, and hope you find BEACHCRAFTS, TOO! an enjoyable follow-up.

To my old friends and my new friends, I say again — Enjoy the book. Enjoy the beach.

Brian Heinz

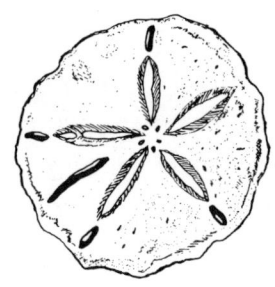

MOCK STONE AGE ART

Maybe you've been to a large museum and seen displays of stone age art left by our early human ancestors on the walls of their caves. It's quite captivating, isn't it?

Early man did not communicate with a formal writing system, as we do today. But these prehistoric paintings were accurate records of life tens of thousands of years ago.

The "paints" of these primitive cave artists were provided by the natural world — dyes of fruits and berries, and the red and yellow pigments of ocher, a soft clay of iron oxide. Or the reddish-brown of umber, from earth containing oxides of manganese and iron. Even the charcoal-black of burnt wood was used.

With these simple elements, early artists left us a pictorial history of their hunting lifestyle and of the animals who shared their environment. Some of the most stunning of these cave paintings, from the paleolithic period, are found in the famous Lascaux cave in the Dordogne region of southwestern France.

You can re-create this art in a most realistic way by following these directions.

Stone Age art requires, first and foremost, a stone! Many beaches abound with large, smooth, light-colored quartz stones. Select one with a surface that is eight to ten inches across. Perhaps you can find one that will sit upright on a table for immediate display when you are done.

Don't select stones of black, grey, or muddy tones. Look for light tones such as white, tan, or pale yellow. Your artwork will stand out more clearly, more dramatically, when completed.

Wash the stone and dry it thoroughly.

To maintain realism, your drawings should portray animals of the period. These include the bison, woolly mammoth, early horse, and the cave bear. A visit to the library will provide books containing pictures of these animals, or perhaps you will copy the Lascaux paintings onto your rock using pictures from reference books.

Your choice of colors should not include "modern" shades of pink, hot orange, or sunshine yellow. Use muted, earthy colors instead. A large box of assorted colored chalks and pastels will provide appropriate colors for authenticity.

Lightly sketch your drawing in pencil. Think primitive! Don't be too detailed and keep freedom in your pencil strokes. Notice how the artwork shown in the photographs (done by a sixth grade girl) seems to flow with the shape of the stone. The bison is clearly in motion, as is the horse, due to the curves and angles captured in the lines.

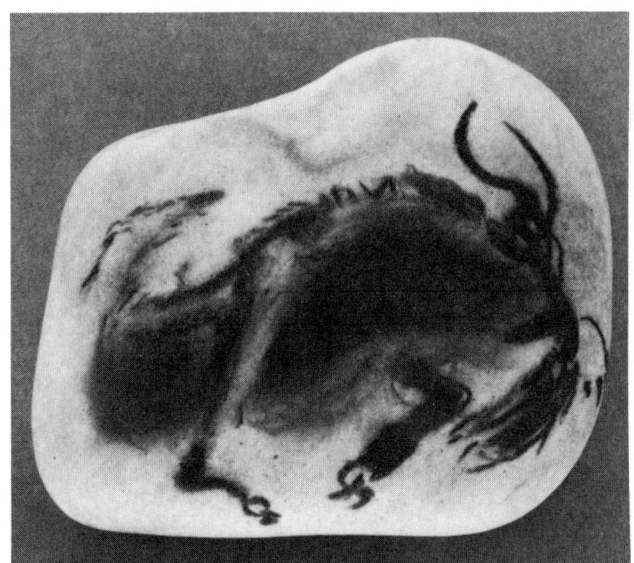

Mock Cave Art by Nicole Michalski, age 12

With your sketch complete, accent your major lines with bold strokes of dark chalk or charcoal. Use reddish-brown or black. Lightly tone in the rest of the picture with soft chalk strokes and rub them gently with your fingertip or a piece of soft tissue to blend and soften the lines and colors. (This is called burnishing.) Highlights, if needed, can be toned in with lighter chalks such as yellowish brown.

At this stage your work is easily smudged. You will protect it with a special spray finish called "art fixative." This is a transparent spray which leaves no gloss, yet will seal your cave painting onto the pores of the stone. This spray is available wherever art supplies are sold. Spray your work with two to three light coats.

Do not use high-gloss products like varnish, shellac, or polyurethane. They produce an artifical look which destroys the realism you hope to capture in your rock art.

Your mock Stone Age art can stand on its own as a work of art, or find a more practical application as a bookend or as a unique paperweight for your desk.

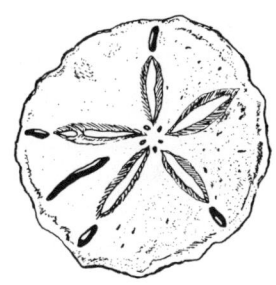

SCALLOP WIND CHIMES

One of the prettier shells on the beach belongs to the scallop, of which there are three common varieties. These include the bay scallop, calico scallop, and deep-sea scallop. Bay scallops are found more commonly by beachwalkers, as bay scallops prefer shallower, inshore waters of eel grass beds. The empty shells are easily thrown ashore by the powerful wave action of heavy storms. Bay scallops have a range extending from the Gulf of Mexico northward to as far as Nova Scotia.

Calico scallops are a more colorful southern species ranging south from Virginia and the Carolinas.

The deep-sea scallop can be found in shallow water (12 feet) from Massachusetts to Labrador, but prefers deeper waters (up to 600 feet) as you move southward to North Carolina.

Bay scallops grow to three inches and are coarsely ribbed with a wide variety of colors and interesting shell markings. Calico scallops are slightly smaller, but brightly colored. The deep-sea scallop grows to eight inches. Its shell is quite flat and smooth with a waxy-white translucent glow, and lacks the pronounced radiating ribs of the bay and calico scallops.

Not only are these animals delicious to eat, but their shells provide a pleasant musical tinkling when assembled as a wind chime. Let's build one!

On your next beach walk, collect as many whole or near whole scallop shells as possible. They can be of varied sizes.

Your shells will probably be bay scallops since deep-sea scallops are only occasionally cast ashore by heavy storms.

Bay scallops are a nice size to work with for this craft, but you may wish to construct a large windchime with deep-sea scallops. If so, you may be able to obtain the shells you need at your local fish market. Ask the owner for the empty shells which are probably discarded after the meat has been removed.

Clean your scallop shells in warm soapy water and rinse them clean. Allow them to dry.

With a light hammer, tap a tiny hole into the rounded crest of the shell using a thin, sharp brad or finishing nail, as shown in the illustration.

Tap hole gently into crest of shell with thin nail and tack hammer.

Lay out your shells in a pattern of four to six columns, depending on how many shells you've collected. Try to have at least four or five shells in each column. Leave one inch of space between the shells from top to bottom, but allow them to overlap slightly from side to side as shown in the diagram.

Cut a stick slightly wider than the top row of scallop shells, so it extends an inch beyond each end. A piece of driftwood, bleached and weathered smooth by the sea and sand, works beautifully.

Insert a screw-eye or cup hook (available at any hardware store) centered above each column of scallop shells. Twist them securely into the wood.

To assemble the shells to the wooden bar and to each other, you'll need a spool of fine plastic fishing line. It is also known as monofilament line. Ask for either six-pound or eight-pound test line. It is strong and nearly invisible, allowing your scallops to appear suspended in air. This type of fishing line can be purchased at sporting goods shops or hardware stores for a modest price.

Cut a length of line for each column of shells. Make the lengths of line twice as long as the columns of scallops to allow for knotting and trimming.

Tie one end of a line securely to the first cup hook. Several overhand knots, pulled snugly, should do it.

Pass the other end through the hole in the first, or topmost, scallop and fasten it in position with a double overhand knot. Pull all knots tight.

Proceed to the next shell, and so on, until the first column of shells is securely fastened to the line.

Finish the remaining columns of shells in the same way. Check your work against the illustration.

Scallop Wind Chimes

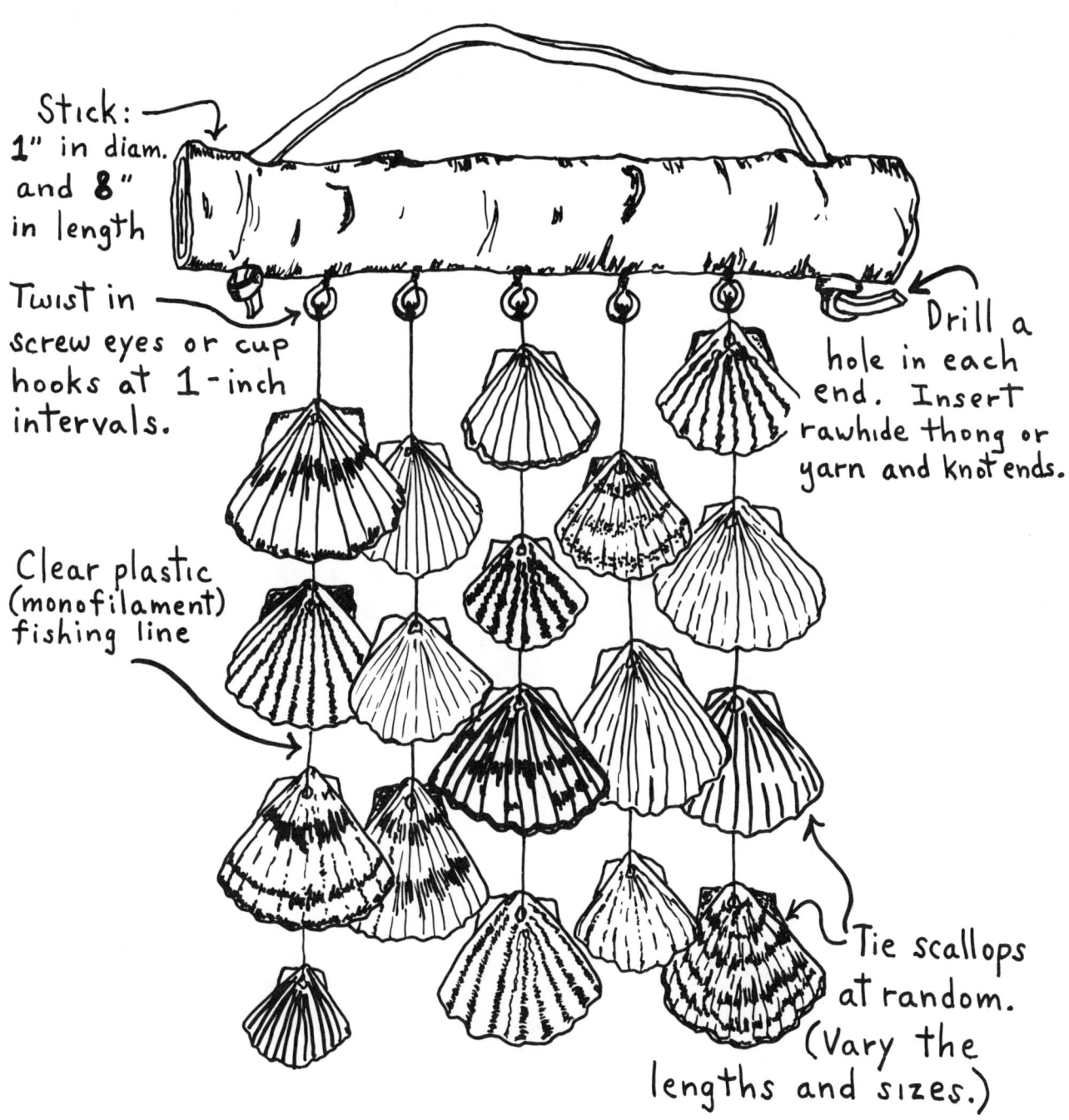

Stick:
1" in diam.
and 8"
in length

Twist in
screw eyes or cup
hooks at 1-inch
intervals.

Drill a
hole in each
end. Insert
rawhide thong or
yarn and knot ends.

Clear plastic
(monofilament)
fishing line

Tie scallops
at random.
(Vary the
lengths and sizes.)

Cut any excess fishing Line from the top and bottom of the wind chime.

Finally, tie a piece of leather thong or braided cord to each end of the mounting stick to use as a hanger, as shown in the illustration. You may want to drill holes in each end of the stick. Insert the ends of the thong and knot the ends to prevent them from slipping through. Or, file a groove around the stick ends and tie the hanging cord around the grooves. This prevents the cord from sliding inward or slipping off the ends.

Hang it outside where it can be blown freely by the wind. Not only does it sound beautiful, but notice how the sunlight glows through the delicate areas of the shells, bringing life to the colors and patterns.

Try this craft with other light shells such as jingle shells. The music created will be quite different.

BEACH WALK A TO Z

A leisurely stroll on the beach at any time of the year can become an activity of interest, an activity to stimulate your thinking and hone your observational skills.

Using the sheet that follows, your purpose is to generate a list of words for each letter of the alphabet and enter those words in the appropriate boxes.

This language exercise develops an area of thought called "fluency," in which your mind works more efficiently and effectively at providing words to a given theme. Your theme here, of course, is the seashore.

Try to consciously remove any barriers to your thinking of words. This means to use all five senses and not to limit yourself only to nouns, (words that name persons, places or things,) but use descriptive words (adjectives), action words (verbs), and include emotional words and slang, or even words that mimic the sounds your hear at the beach such as *"squish," "plop,"* and *"whoosh."*

Here are examples for the letter A: *airy, animal, algae, ah-h.*

If your feet are tired or sore from walking, feel free to use words like *ache* and *aggravation.*

This activity is also fun to play in pairs or teams of friends. Start with A and take turns naming words until one player is stumped. The last person to provide an acceptable word for A gets a point. The game proceeds to letter B, C, and so on until the alphabet is complete. The person with the greatest number of points at the end is the winner. And you can celebrate your victory over a salty bowl of alphabet soup.

BEACH WALK A to Z

A		**N**	
B		**O**	
C		**P**	
D		**Q**	
E		**R**	
F		**S**	
G		**T**	
H		**U**	
I		**V**	
J		**W**	
K		**X**	
L		**Y**	
M		**Z**	

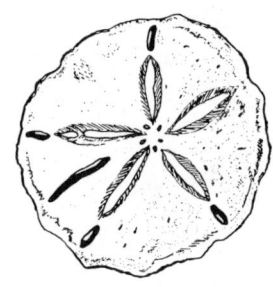

SHELL MIRRORS AND PHOTO FRAMES

Here's a "picture perfect" idea to change a jumbled bunch of seashells into an artistic assemblage worthy of framing. Come to think of it,...it *is* a frame, suitable for photographs or a mirror.

To begin, you'll need a basic wooden frame. These can be purchased very inexpensively at craft shops, photography stores, or department stores. They are usually unfinished pine and come in standard rectangular or oval sizes of five by seven, eight by ten, and larger. Old wooden frames can often be purchased at garage sales for pocket change.

Or, an inexpensive (cheap) frame can be easily made using wooden canvas stretcher strips purchased at any art supply shop. These strips are used by artists to assemble frames upon which they secure their canvas in preparation for painting. The strips come in assorted lengths and have pre-mitered joints at each end. The various lengths of strips can be combined to produce many rectangular sizes, since all the joints are grooved to slide firmly into one another without glue. Glue them anyway with white glue.

Once you've decided upon your frame, you'll need to gather your shells. Remember, the larger the frame, the greater the number of shells you'll need to cover the surface of the wood.

You need not worry about collecting perfect shells. Incomplete shells and shell fragments worn smooth by the sand and sea can be used quite successfully, since many of the shells will overlap each other. Stick to using smaller shells. Large shells tend to overpower the frame and look bulky and out of place.

Wash all the shells in warm soapy water. Remove all loose sand and surface grime. Rinse them well and let them dry thoroughly.

Shell Mirror or Photo Frame

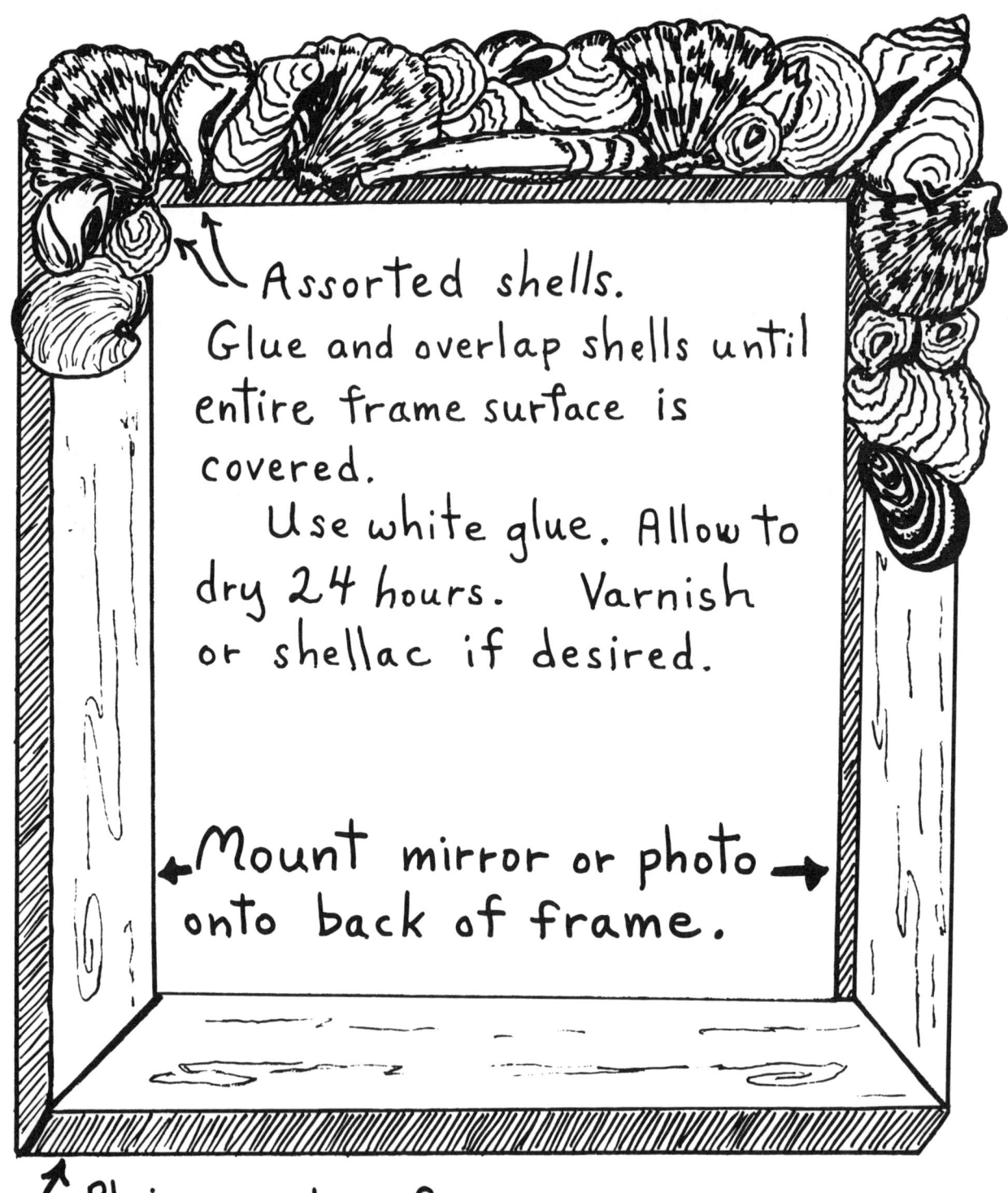

Assorted shells.
Glue and overlap shells until entire frame surface is covered.

Use white glue. Allow to dry 24 hours. Varnish or shellac if desired.

←Mount mirror or photo→ onto back of frame.

↙Plain wooden frame

Using white household glue, begin mounting the shells onto the frame. Begin at one corner and proceed in one direction around the frame. Do not work in several areas at once and do not try to proceed too quickly. Allow a section to dry for a day before continuing. Be sure your frame is placed where it will not be disturbed while the glue sets.

Even though white glue dries clear, don't use it too heavily. If it runs or spreads onto the shell surfaces, it will detract from the natural beauty and reduce the luster. The glue is quite strong. Use just enough to secure the hidden surfaces and edges of one shell to another, or to the frame.

Before shells are actually glued, lay them out along the frame. Rotate them, exchange their positions, vary the colors and shapes as you go. When you're satisfied with the arrangement, begin to glue.

When the frame is completely covered with glued shells, allow your work to dry for a full day.

As a final step, give the shell frame several thin coats of spray varnish or shellac. Allow each coat to dry for thirty minutes before applying the next coat. This will produce a brilliant shine and deepen the natural colors of the shells.

Your frame is now ready to receive your favorite photograph or a mirror. (Mirrors can be cut to the frame size at your local glass shop.)

Whatever you decide, your shell frame will be a reflection of your talent and will grab the attention of your friends. They make unique personal gifts, too. Think about it. And here's looking at you, kid!

DICTIONARY SCAVENGER HUNT

Nature has provided us with an almost infinite world of colors, aromas, sights, sounds, shapes and textures. An artist attempts to capture most of this with paints on canvas. But what about a writer? The myriad details of the outdoors can be captured by writers only when they have just the right words. In fact, one could spend a thousand words in describing the dirt beneath your feet.

The dictionary scavenger hunt is a great game to develop a more powerful vocabulary, sharpen your observational skills, and become proficient in the use of a dictionary while getting nose-to-nose with the natural world.

Twenty-four words are provided on the accompanying page. You'll need a decent paperback dictionary, suitable for field use, and a plastic collecting bag.

If you choose to do this activity alone, purely for enjoyment and learning, you need not be concerned with time. Work at your own pace. It becomes more exciting, however, when teams of two compete against one another with a time limit imposed.

To play, the teams go to the selected site. Perhaps a field or marsh is chosen, a local woodland, or the beach. A time limit is agreed upon of ten to twenty minutes. At the word GO, the teams head out and begin collecting items from the environment described by each word from the list. Certain words on the list may already be familiar to some players, but other words must be looked up and defined before an object is collected.

Use restraint and common sense in collecting any live plants or small animals (caterpillars, for example) unless they will be returned immediately to the environment without harm following the game. If the leaf of a certain plant is needed to illustrate a word, remove only one leaf. Do not uproot the entire plant.

At the end of the time limit, finished or not, the teams return to the starting area where points are awarded. A team must produce an object and verify to the others that it conforms to the definition of a particular word on the list. Sometimes a team finds an item that fits several different words on the list at once. This can provide the fuel for some fiery arguments! Decide ahead of time if multiple words for one object will be accepted. Award one point for each correct word/object match. The team with the most points wins.

As an extension, mount your findings neatly on plywood or oaktag with white glue. Then cut out the words and paste them down as labels beneath your objects.

Dictionary Scavenger Hunt

Use a dictionary to define each adjective. Find beach items described by each word and glue them neatly on cardboard. Cut out adjectives and paste as labels beneath each item.

GRANULAR	SYMMETRICAL
SYNTHETIC	EDIBLE
CRIMSON	COARSE
CIRCULAR	FRAGILE
FLEXIBLE	VELVETY
HOLLOW	ABRASIVE
METALLIC	CYLINDRICAL
ELASTIC	BRITTLE
TRANSPARENT	ORGANIC
POROUS	BUOYANT
DELICATE	ELLIPTICAL
OPAQUE	LUSTROUS

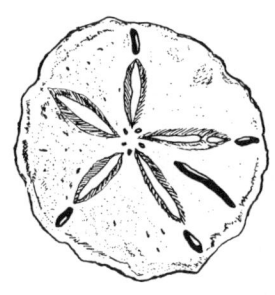

MESSAGE IN A BOTTLE

Ahoy, mates! Every landlubber has seen their share of pirate movies. And how some poor wretch is marooned on a desert isle with no way to send for help other than an empty rum bottle. He scrawls a message on a paper scrap, seals it tightly inside the bottle with a cork, and casts it into the sea. An old idea? Yes. But it can work. And getting a response to your message can be quite exciting. You might even learn something about the natural ocean currents of your area. The trick is to ensure the bottle drifts to another location, and is not simply tossed ashore at your feet again by the next wave. Just follow these steps.

The Message

Of course, you can write any message you please, silly or serious, but why not make a scientific investigation out of this activity? In your message, state the exact date, time, and place that you launched your bottle. Include some personal information about yourself as well. You can be sure the finder of your bottle will be a bit curious about you. If you place a stamped postcard or envelope into your bottle, it may motivate the finder to respond more quickly. Ask them to tell you a little about themselves and to record the date, time, and place where your message was found.

With this information, you can check a coastal map of your area and trace the route your bottle traveled. This will be a key to the drift or current pattern in your area. Also, determine the approximate mileage of your bottle's voyage by using the map's mileage scale. Assuming your bottle did not remain undiscovered for too long before being found, you can even compute the bottle's speed or rate of travel.

Here's an example.

You launched your bottle from your hometown at 9:30 in the morning on May 10th. It is reported found at 2:30 in the afternoon on May 12th at a town down the coast.

The total time of travel was 52 hours. The mileage between the two towns acccording to the map is 65 miles. Divide the total number of hours into the total miles traveled, as shown below, to find the rate of speed.

$$
\begin{array}{r}
1.25 \text{ miles per hour or } 1\frac{1}{4} \text{ miles per hour} \\
52 \overline{)\ 65.00} \\
-52. \quad\ \\
\overline{130} \\
-104 \\
\overline{260} \\
-260 \\
\overline{0}
\end{array}
$$

A Proper Bottle Launch

With your message written and ready, it's time to properly prepare your bottle for the launch.

Choose a clean one-liter or two-liter soda bottle with a tight-fitting screw cap to transport your message. The plastic bottle is more resilient than glass and certainly less of a hazard should it break on the rocks of a local shore.

The trick is to get your bottle to float upright in the water with the neck and cap just above the surface, as shown in the illustration. This is accomplished in the same fashion that a ship is given stability in the water, by adding ballast, or weight, to the base of the hull which lowers the center of gravity.

If this is not done, the bottle will ride too high on the water's surface and be blown and tossed about by the wind and waves. You want your bottle to drift with the current.

Your ballast is readily available at the beach in the form of pebbles. (Sand will also work.) Scoop up a handful of pebbles and pour them into the bottle. Screw on the cap and place the bottle into the water. Does it float straight up and down with the cap just above the water's surface? If not, add more pebbles and test it again, until the bottle floats as described. (See the illustration.) If the bottle sinks beneath the surface, obviously some pebbles must be removed.

Message in a Bottle

1 or 2 liter soda bottle

Unweighted bottle floats too high upon water's surface and is easily affected by the wind and waves, causing it to return to shore.

Weighted bottle floats upright below the surface and is not affected by the wind, but carried by current.

⌐ beach pebbles used as ballast.

    With your bottle adjusted, roll up your message and insert it into the bottle. Screw on the cap tightly and launch your bottle by wading into the water, if possible, and tossing your plastic vessel beyond the breakers of the shore. You can also drop your bottle into the water from a pier or from a boat while fishing or sailing. If necessary, you can decrease the chance of water sneaking into your bottle by wrapping a layer of plastic wrap over the mouth of the bottle before screwing on the cap. Or, make several turns around the cap with cloth duct tape. This tape comes in vivid colors and may increase the chances of your bottle being spotted and recovered by someone. The tape is available at any hardware store.

    That's it. Except for the waiting, of course.

    Yo-ho-ho and a bottle of...?

WILY WEB OF WRIGGLING EELS
WORD SEARCH

The tangled mass of eels on the following page provides the grid for this slippery maritime puzzle. Hidden within the bodies of the sixteen hungry eels are thirty-three items commonly found at the shore or in the sea.

Words run horizontally and vertically only (not diagonally) but may run either forwards or backwards. The master word list appears below, but see how many words you can discover on your own before going to the list.

The solution is not provided, so be sure none of the words slip away from you. Good luck!

barnacles	flounder	seahorse
fiddlers	swan	skate
kelp	scallop	cod
sandworm	tuna	grass
shrimp	mussel	quahog
fluke	clam	conch
reeds	bather	tern
slipper shell	oyster	gull
bait	whelk	snail
manta	urchin	salt
snapper	lobster	periwinkle

Wily Web of Wriggling Eels Word Search

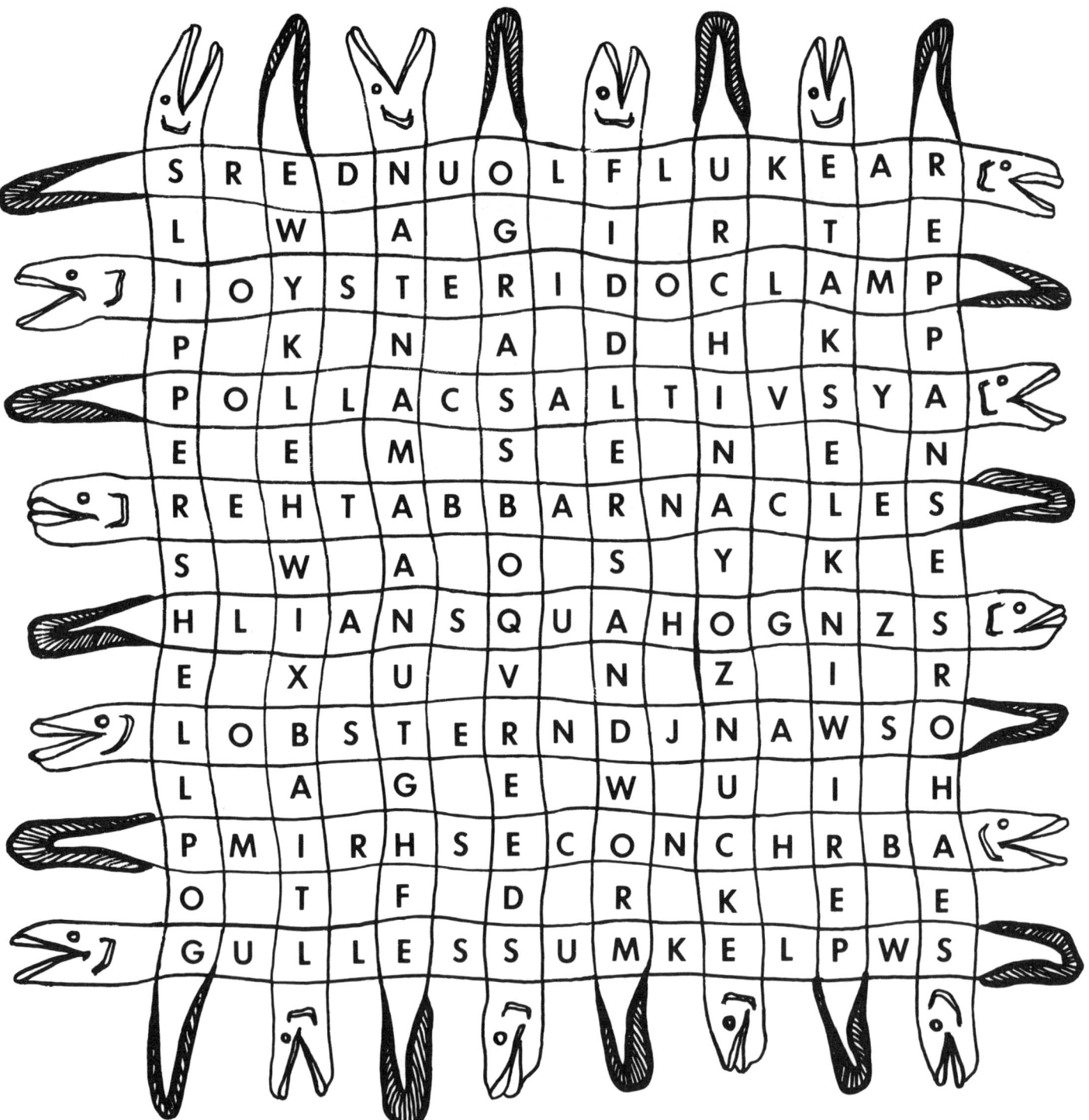

```
S R E D N U O L F L U K E A R
  L W A G I R T E
  I O Y S T E R I D O C L A M P
  P K N A D H K P
  P O L L A C S A L T I V S Y A
  E E M S E N E N
  R E H T A B B A R N A C L E S
  S W A O S Y K E
  H L I A N S Q U A H O G N Z S
  E X U V N Z I R
  L O B S T E R N D J N A W S O
  L A G E W U I H
  P M I R H S E C O N C H R B A
  O T F D R K E E
  G U L L E S S U M K E L P W S
```

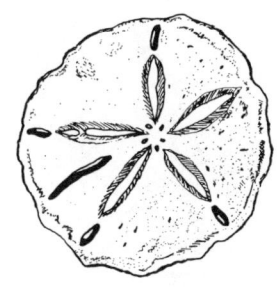

ROCK! ROCK! WHO HAS MY ROCK?

How good is your sense of touch? Can you identify things in the dark simply by feeling them? Want to find out? Then this sensory game is for you.

Next time you're at the beach with a group of friends, ask each person to find a "favorite" rock that can be held in the palm of their hand. The larger the group of people, the more fun the game.

When everyone has selected a stone, have them all sit in a circle, shoulder to shoulder, with their backs to the center of the circle.

Ask the group to close their eyes. Give them a minute to feel their stones until they're reasonably sure they can identify their personal rock by touch only. Feel the texture. Feel the weight. Feel the length, the width, and the height. Does it have any edges? Or any dimples?

Now collect everyone's stones and mix them up in the circle's center. Be sure all players still have their eyes tightly closed. Distribute the rocks at random, placing a stone in each person's hands. Each player examines the stone given to them using only the sense of touch. Does anyone have their personal rock? If not, each person passes the stone to the next person on their right.

When a player believes they have their own original rock, they may call out, "Rock! Rock! I have my rock!" and open their eyes. If they're correct, they may keep their eyes open and remove their rock from the game, but they remain in the circle to continue passing stones until each player has correctly identified their personal rock. If a person incorrectly guesses the identity of their stone, they simply close their eyes again and continue in the game.

There are ways to make the game more interesting and challenging. For one, set strict size limits on the stones. For example, stones must be no more than four finger-widths long and no less than three finger-widths wide. For another, remove one or two of the stones selected by the players and replace them with other stones. (You decide whether or not you want the players to know about this little deception of "red herrings" before the game begins.

19

Continue to pass stones to the right (clockwise) until each player has identified their personal stone.

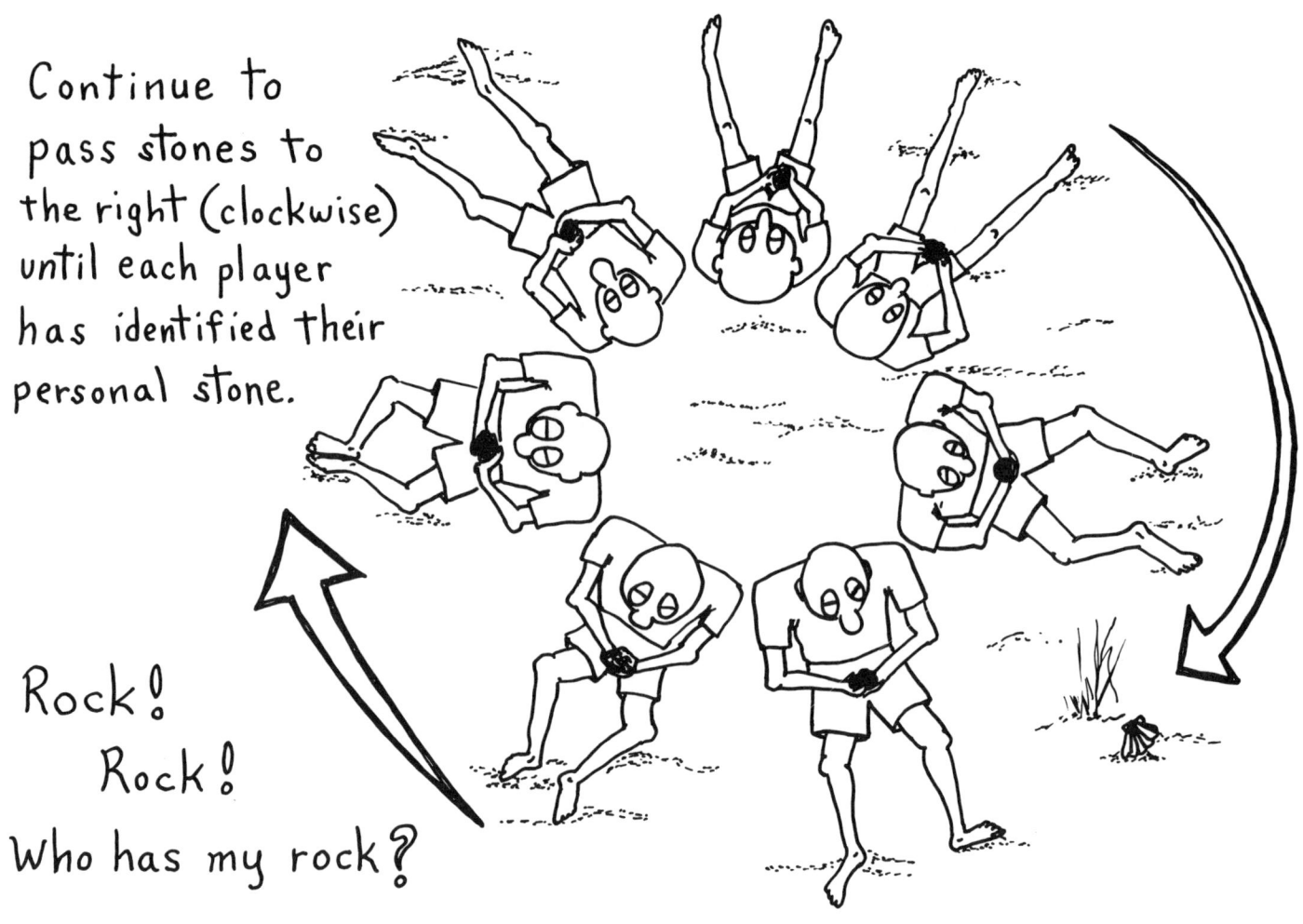

Rock!
    Rock!
Who has my rock?

Most beach stones tend to be fairly rounded and smooth, so it will require sensitive fingers and clear memories to identify the stones correctly.

SEASIDE PICTURE BINGO

The old game of bingo has been given a new twist. But you won't need bingo chips. A pencil and a pair of sharp eyes will do. The grid of twenty-five boxes on the next page contains the names and illustrations of objects and forms of life common to most beaches and shore areas.

A beach is never quite the same place two days in a row. Waves continue to wash new objects ashore while washing other materials back out to sea. Visitations by animals and people also change the nature of the beach and what you can expect to find there on a day-to-day basis. For that reason, the picture bingo sheet will produce different results and different winning combinations at various times of the year and in various locales.

As in the standard game of bingo, the purpose is to see and record five items across, up and down, or diagonally. Keep the game enjoyable with a group by awarding a "win" to each new row or column completed. The grand champion is the first person to complete all twenty-five boxes.

A field guide can be a valuable asset to this game. (Try the Golden Guide to the Seashore or other guides listed in the bibliography at the end of the book.)

Shore areas are divided into zones: surf, lower beach, upper beach, primary dunes, swale, secondary dunes, and possibly a marsh, bay, or woodland. A bit of simple research will guide you to the best location, or zone, to find the plants and animals illustrated in the grid.

Field guides often contain clear photos and color illustrations, along with explanatory paragraphs to further aid you in your discoveries.

You may wish to collect certain items for later use in completing some of the arts and crafts found in this book. If so, bring along a collection bag. Happy hunting!

SEASIDE PICTURE BINGO

Sponge Seaweed (Codium)	Jellyfish	Starfish	Sea Lettuce (Ulva)	Sand Dune
Eel Grass	Hard Shell Clam (Quahog)	Dune Grass	Clouds	Scallop
Heavy Waves	Gull Feathers	Herring Gull	Driftwood	Reeds (Phragmites)
Whelk Shell	A Shore Bird's Tracks	Razor Clam	Oyster	An Animal Bone
Black Stones	Old Fishing Line	Whelk Egg Cases	Tube Worms On a Stone	Acorn Barnacles

22

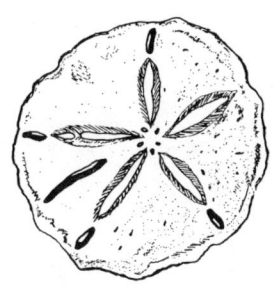

BEACH PEBBLE POLISHING

We've all shared the experience of walking a beach and having our attention drawn suddenly to the ground by a stone with a fiery color or unique pattern. We're forced to pick it up and examine it more closely. We wonder about its origins within the earth. Often, it gets dropped into a pocket for saving. These "special" stones, and most other stones on the beach, have the potential of becoming gleaming beauties that can rival many gemstones.

Even drab and ordinary stones and pebbles can be given a deep, brilliant, polished surface that will reveal subtle hues and fascinating grain patterns that were not at first evident. All you need do is collect your favorite pebbles. The real work is done by a simple machine called a rock tumbler.

The device is nothing more than a small drum, or barrel, with a tight-fitting, removable cap. The drum contains your assortment of stones, some water, and the polishing grits. A small electric motor rotates the barrel on two rollers, continuously tumbling the stones through the water-grit mix, called "slurry," until a finely polished surface is achieved. This requires some patience, as the process takes three to four weeks. But the results are more than worth the wait.

Your tumbler requires four grades of abrasives. Each is used individually as the stones are tumbled through increasingly finer grades. The abrasives are usually labeled coarse, medium, fine, and polish, or coarse, fine, pre-polish, and polish.

Beginning with the coarse grit, your tumbler will run twenty-four hours a day for a week. At the end of the week, the stones are removed and rinsed clean along with the tumbler barrel. Then the stones are returned to the barrel and the medium grit is added. Continue the procedure with the fine and polish grades of grit.

A Simple Tumble Polishing Machine

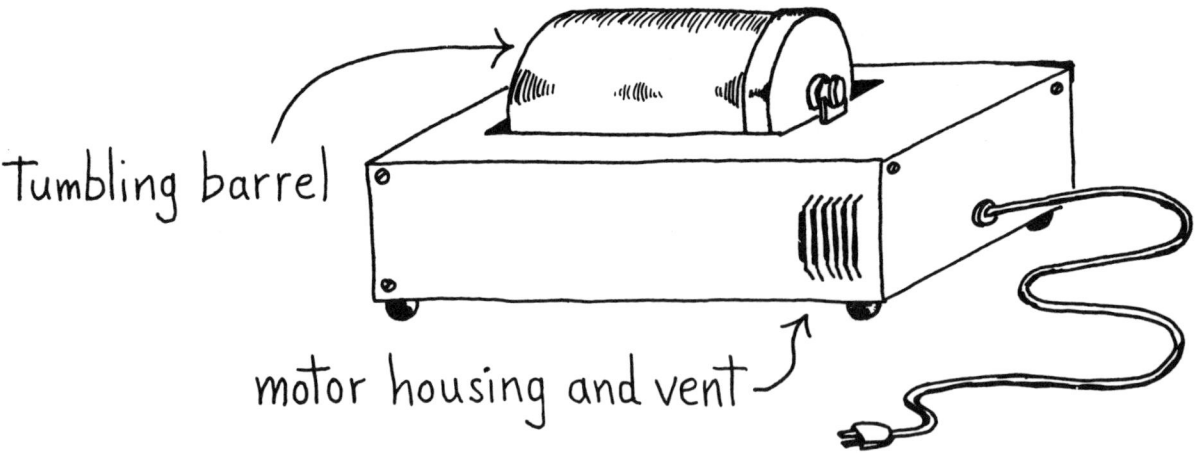

tumbling barrel

motor housing and vent

Be sure *not* to rinse your used slurry into your sink drain. It will harden like concrete and clog your pipes. Rinse the barrel and stones outdoors with a bucket and hose.

There is nothing as beautiful as a ceramic bowl piled high with these colorful, sparkling geological beauties displayed in your den or living room. They can be given as gifts. Many smaller beach pebbles, when polished, resemble jelly beans in soft pastels. I've seen tiny woven baskets of them given away as Easter gifts. Be sure your guests are told that these look-alike jelly beans are, indeed, stones!

Where Do I Get A Rock Tumbler?

Inexpensive rock tumblers can be purchased at your local lapidary. These shops specialize in jewelry, artwork, and collectible specimens of gemstones, minerals, fossils, shells, and other natural and geological resources.

Often, high quality toy stores and hobby shops that carry a line of science and craft kits also carry tumblers.

If you are unable to locate an appropriate shop nearby, even after checking your telephone business directory, try contacting the shops listed below.

Ecolin Co., Inc.
Lighthouse Landing
14 East Broadway
Port Jefferson, NY 11777
(516) 473-1117

This store is a geological wonderland of gems, metals, crystals, minerals, fossils, and other earthly delights. The owners of Ecolin, Russ and Linda Baker, carry a variety of tumbler sizes to choose from. The smaller ones will polish up to six pounds of stones and are quite reasonably priced. Pre-packaged polishing grits are also available. Call them for further information and prices.

Copernicus
394 New York Avenue
Huntington Village, NY 11743
(516) 271-7919

Copernicus
134 Main Street
Port Jefferson, NY 11777
(516) 473-5450

Both of the Copernicus shops specialize in scientific toys, games, equipment, and kits. By calling or writing to the owners, Harris and Ellen Tobias, you can receive their catalog. They do carry rock tumblers.

The Village Toy Shop, Inc.
55 Main Street
East Hampton, NY 11937
(516) 324-6455

This shop, located on the east end of Long Island, also carries a wide array of scientific toys which includes rock tumblers. Contact the owner, Rich Wilson. He'll be glad to help.

So, the next time you find yourself traveling a rocky road, pick some up and smooth things out a little with your tumbler. The result will be a shining example of your creativity.

THE GREAT MUD SNAIL RACE

"Gentlemen, start your engines!" is the traditional command heard over the loudspeakers at a famous auto race.

Everyone loves an exciting competition of speed. Well, here's a more quiet race, without the deafening roar of engines, that can be just as exciting. And it all takes place in a cake pan!

Your race vehicles will be living creatures commonly called mud snails, although they are more correctly known as mud dog whelks (*Nassarius obsoletus*). These interesting little snails grow to only one inch and are collected easily in very shallow intertidal waters at the edge of quiet bays or in the salt creeks of marshes. They are dull and dark in color and prefer soft, muddy bottoms. Often, they congregate in large numbers in only several inches of water and can be seen easily below the water's surface.

Collect about a dozen mud snails and place them in a shallow bake pan that is about twelve inches wide and eighteen to twenty-four inches long. Cover the bottom of the pan with a thin, even layer of sand from the bayside and fill the pan with an inch or two of sea water before adding the snails. The layer of sand will show clearly the paths of travel followed by each snail.

In order to "run" the best race, there must be a prize offered to the contestants at the finish of the race. For people, the prize may be a trophy, money, or simply an honor. For mud snails, the prize must be food.

Mud snails are meat-eating (carnivorous) scavengers with a well-developed sense of smell. They will detect and feed on any dead animal material on the bay bottom.

Place your mud snails at one end of the cake pan. Then place a cracked mussel, clam, or piece of bait fish at the other end. Let the race begin! Mud snails can travel about two inches per minute, so your race should not last much more than ten minutes.

The Great Mud Snail Race!

Flat, shallow pan with thin layer of sand and 2 inches of water

Bait (cracked mussel) at this end

Release your mud snails (mud dog whelks) at this end.

At the conclusion of the race, mud snails will be swarming all over the bait, feeding on it with a tube-like extension called a proboscis. The proboscis contains the radula, a rasping tongue, that shreds away the meat to be eaten.

To make the race even more exciting, get some friends to attend your race. Let each person select a snail as his or her "race car." You can even mark the shells for easy identification. Remove the mud snails from the tray and gently wipe dry the upper surface of the shells. (The snails will withdraw into their shells while you do this.) Next, put a tiny drop of nail polish on the shell and blow on it for a moment until it is dry. Use a different color for each snail, or paint a tiny letter on the snail shell with the nail polish brush.

Run the race several times. Does the same snail win each time? Do the snails travel in a straight path? If not, why? How long does it take for them to begin sensing the food? How much can you learn about the behavior of these creatures?

You can run your races right on the beach and release the snails when you are done. Or, keep the snails in a simple salt-water aquarium to run the races whenever you wish. These snails are quite hardy and do well in marine aquariums.

Who knows? You may just find the fastest mud snail on earth!

OBSERVATION WALK SHEETS:
Ocean, Swale, and Bayside

Learning about the interesting, colorful, and abundant forms of life at the shore can be as simple as taking a walk. You need only a pencil and clipboard, the observation walk sheets on the following pages, and a good field guide. (Try *The Outer Lands* by Dorothy Sterling or *A Field Guide to the Atlantic Seashore* by Kenneth L. Gosner.) These guides and others are listed in the bibliography. You'll find them helpful, especially for the illustrations they provide of plants and animals with which you may be unfamiliar. Certain terms on the observation walk sheets may be new to you. If so, turn to the index of your field guide, look up the word in question, and flip to the indicated page.

Separate sheets are provided for the ocean, the swale, and the bayside. Each area is unique for its treasures. The following tips may be helpful.

Ocean - Believe it or not, the best times for finding evidence of sea life on the ocean beach are late fall, winter, and early spring. Human traffic is scarce and the heavier winds and wave action churn up debris from the sea floor and cast it ashore. Dress warmly.

Swale - Give special attention to the swale area. The many fruit-bearing plants that grow here provide important food sources for mammals and birds throughout the year. These plants include catbrier vines, beach plums, salt-spray rose, blueberries, wild cherry, poison ivy, and bayberry.

Bayside - The shallow waters of the bay are often fringed by reeds (*Phragmites communis*) or marsh grasses like cord grass (*Spartina alterniflora*) and salt-meadow grass (*Spartina patens*). These are ideal areas to observe marsh birdlife.

Snails thrive here, too. Mud snails are carnivorous scavengers often found at the water's edge. Periwinkles are vegetarian snails, climbing on exposed rock surfaces to scrape away the algae with their rasping tongues, or radulas. Dip nets or seine nets are an asset in seeking bay life. Many forms hide in the dense eel grass. Helpful ideas on constructing and using these nets are found in my earlier text, *Beachcraft Bonanza*. Also, see the related unit in this text, "The Amazing Mud Snail Race."

The observation walk sheet activity is a wonderful way to spend some quiet time at the shore with a close friend, casually discovering the marine environment. It's also a highly productive field trip activity for youth groups, as it re-enforces research skills.

None of the lists is, by any means, complete. Your particular geographical area and the time of year you select to try this activity will have a direct influence on what you will see. Many forms of plant and animal life may be observed which are not stated on the lists. Blank lines have been provided for you to add these species.

OCEAN OBSERVATION WALK

Look for these items in part or whole on the beach or on the dunes at the oceanside. Consult your field guides for help and pictures. Check off the circle for each item found. Use the blank lines to record specimens not listed here.

○ Rockweeds (*Fucus*)
○ Sea Lettuce (*Ulva*)
○ Scallop Shells
○ Starfish
○ Hermit Crabs
○ Kelps (*Laminaria*)
○ Sulphur Sponge
○ Knotted Wrack
○ Mermaid's Tresses (*Cladophora*)
○ Horseshoe Crab
○ Quartz
○ Deadman's Fingers Sponge
○ Irish Moss (*Chondrus crispus*)

○ Channeled Whelk
○ Mole Crab
○ Slipper Shell
○ Oyster Drill
○ Blue Crab
○ Seaside Goldenrod
○ Skate Egg Case
○ Knobbed Whelk
○ Surf Clam
○ Jingle Shell
○ Dusty Miller
○ Moon Snail
○ Common Tern

○ Rock Crab		○ Sand Collar (moon snail eggcase)	
○ Sponge Seaweed (*Codium*)		○ Blue Mussel	
○ Spider Crab		○ Beach Pea	
○ Shipworms (*Teredo*)		○ Magnetite	
○ Ribbed Mussel		○ Red Garnet Sand	
○ Dune Grass		○ Sea Rocket	
○ Herring Gull		○ Driftwood	
○ Loon		○ Beach Glass	
○ Seaside Spurge		○ Canada Geese	
○ Great Black-Backed Gull		○ Seabeach Orach	
○ Gooseneck Barnacles		○ Laughing Gull	
○ Cormorant		○ Bufflehead	
○ Whelk Egg Case		○ Angel Wing Shell	
○ Acorn Barnacles		○ Chenille Weed (*Dasya*)	
○ Black Duck		○ Sandpiper	
○ _____		○ _____	
○ _____		○ _____	
○ _____		○ _____	
○ _____		○ _____	
○ _____		○ _____	
○ _____		○ _____	
○ _____		○ _____	

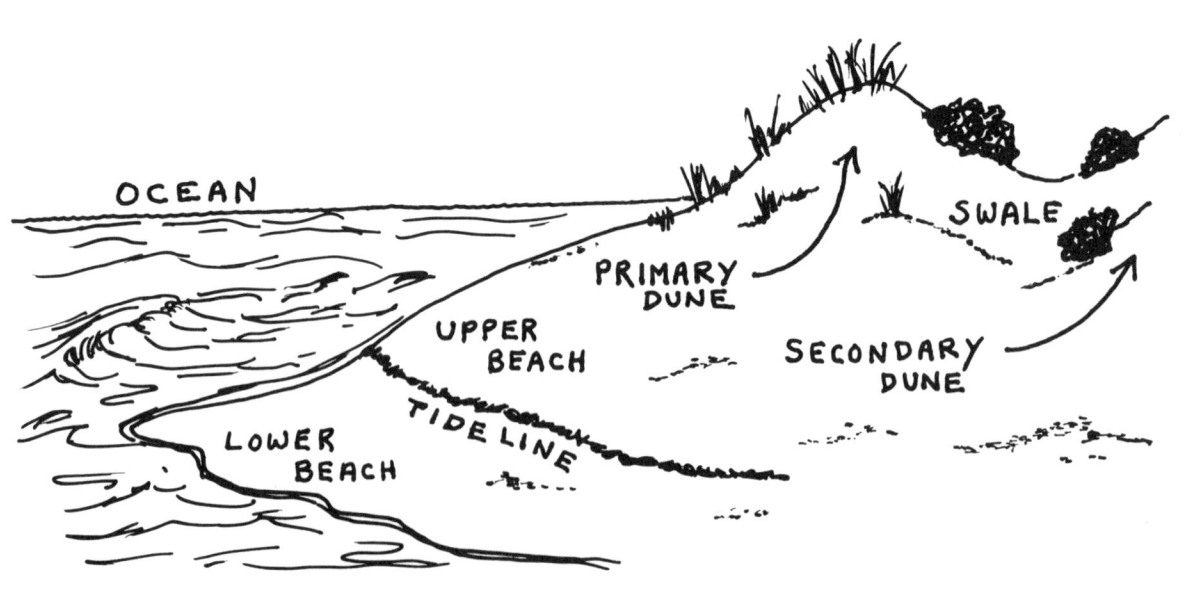

OCEAN

PRIMARY DUNE

SWALE

UPPER BEACH

SECONDARY DUNE

TIDE LINE

LOWER BEACH

SWALE OBSERVATION WALK

Look for these items in the swale, the sheltered area behind or between the dunes. Consult your field guides for help and pictures. Check off the circle for each item found. Use the blank lines to record specimens not listed here.

○ Beach Plum

○ Mourning Dove

○ Poison Ivy

○ Sparrow Hawk

○ Bearberry

○ Red Fox

○ Short-tailed Shrew

○ Golden Aster

○ Seaside Goldenrod

○ Wild Cherry

○ Virginia Creeper

○ Red Chokeberry

○ Yarrow

○ Song Sparrow

○ Warblers (varied)

○ Prickly Pear Cactus

○ _____

○ _____

○ _____

○ _____

○ _____

○ _____

○ Bayberry

○ Tree Swallow

○ Beach Heather (*Hudsonia*)

○ Marsh Hawk

○ Rabbit

○ Deer

○ Red-winged Blackbird

○ Beach Grass

○ Highbush Blueberry

○ Pitch Pine

○ Cat Brier

○ Wild Grape

○ Salt-spray Rose (*Rosa rugosa*)

○ Blackberry

○ Reindeer Moss (*lichen*)

○ Earth Star (*fungus*)

○ _____

○ _____

○ _____

○ _____

○ _____

○ _____

BAYSIDE OBSERVATION WALK

Look for these items along the bayside and marsh. Consult your field guides for help and pictures. Check off the circle for each item found. Use the blank lines to record specimens not listed here. (You may need a net.)

○ Reeds (*Phragmites*)
○ Quahog (hard-shell clam)
○ Bay Scallop
○ Spider Crab
○ Hermit Crab
○ Marsh Hawk
○ Mud Snail
○ Acorn Barnacles
○ Fiddler Crab
○ Cord Grass
○ Red-Winged Blackbird
○ Groundsel
○ Blue Mussel
○ Razor Clam
○ Black-crowned Night Heron
○ Atlantic Pipefish
○ _____
○ _____
○ _____
○ _____
○ _____
○ _____

○ Marsh Elder
○ Soft-shell Clam
○ Killifish
○ Eel Grass
○ Evening Primrose
○ Periwinkle
○ Grass Shrimp
○ Sea Blite
○ Osprey (fish hawk)
○ Salt Meadow Hay
○ Glasswort
○ Marsh Elder
○ Ribbed Mussel
○ Tube Worm
○ Greater Yellowlegs
○ Yellow-crowned Night Heron
○ _____
○ _____
○ _____
○ _____
○ _____
○ _____

It is not necessary to complete all three sheets in one visit. Rather, familiarize yourself slowly with one area, learning a few species at a time. Before long you will become an authority on the most common forms of shore life in your area.

But, be prepared. Sooner or later something most unexpected will show up,...a new mystery to identify and add to your seashore lore.

SAND COMBS AND COMBING

No one at the beach can resist playing with, admiring, or otherwise enjoying the textures of the sands. People stretch out, face down on blankets, and wiggle their toes in it. Others sift it gently between their fingers. Still others admire the beautifully textured patterns created in the sand by nature; the gently scalloped terraces left by the wind, or delicate swirls etched by the brushing tips of dune grasses, and the firm smooth bands left by retreating waves.

Sand combing provides the pleasure of playing with sand while creating your own original patterns and textures.

A sand comb is a simple tool to construct. Start with a piece of heavy, corrugated box cardboard or plywood. Cut a rectangle of six to twelve inches in length and four to six inches high, as shown in the illustration.

Draw an interesting pattern of "teeth" on each edge to produce a double-sided sand comb. Make each pattern different. Use simple geometric designs such as semi-circles, squares, or triangles. Carefully cut out the teeth.

To use the comb, press the teeth into the sand, but not beyond the depth of your comb's teeth. Draw the comb smoothly through the sand. Work in dry sand. Work in wet sand. Which produces better results? Try combining several combs for more intricate and interesting patterns and textures.

Sand Combing

Draw an interesting pattern of "teeth" onto a piece of thin wood or heavy box cardboard, 6 to 12 inches long and 4 inches high.   Cut out your sand comb.

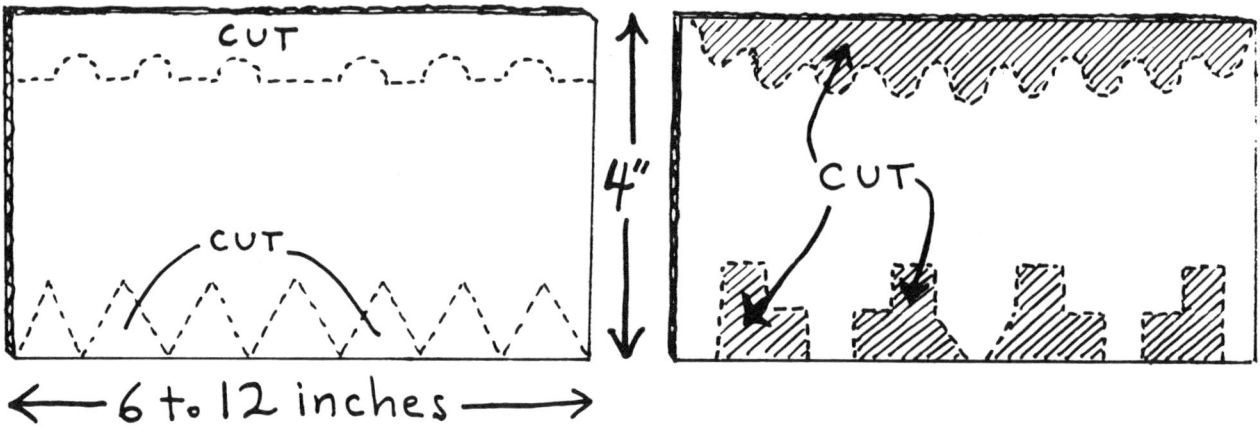

Make your designs simple or complex.   Cut teeth patterns into both sides of your sand comb and alternate your designs by turning over the comb.

Pull your combs gently through the sand. Try it in wet sand. Try it in dry sand. And try criss~ crossing your different patterns.

DROPS OF SWIRLING SEA WATER EXPERIMENT

If someone asked you, "What's the difference between sea water and the water from your faucet, or tap water?" you might say, "That's easy. Sea water is salty." And, of course, you would be right. But it's not the only difference. Did you know that sea water is heavier than fresh water?

Equal volumes of fresh water and sea water do not weigh the same. The dissolved salts and minerals in the ocean cause it to be more dense than fresh water. You could prove this by measuring equal jars of tap water and ocean water and weighing each on an accurate, sensitive scale. But there is an easier and more dramatic method. Here's how.

Collect a jar of sea water and a jar of fresh water. Get four small, clear glasses. Plain glasses without designs are best. Fill two glasses with sea water and the other two with fresh water. Mark the glasses with an "F" for fresh water and an "S" for salt water. (See the illustration.)

Place one glass of fresh water next to a glass of salt water and get an eyedropper. Fill the eyedropper with salt water and hold the tip of the dropper about one inch above the glass of fresh water. Look closely into the glass at the water and slowly release one single drop of salt water into the glass of fresh water. Be sure to squeeze the eyedropper bulb gently. What did you see? Did the drop become visible as a sort of swirl? Did it plunge right to the bottom? Try it again. Observe it closely. Try it once more.

The swirling lines that you see are caused by the difference in density between salt water and fresh water. Wherever and whenever the two different waters meet, they form an *interface* and you see wavy lines as a result.

The fact that the droplet of salt water falls quickly to the bottom of the fresh water proves that the salt water is heavier.

Drops of Swirling Sea Water Experiment

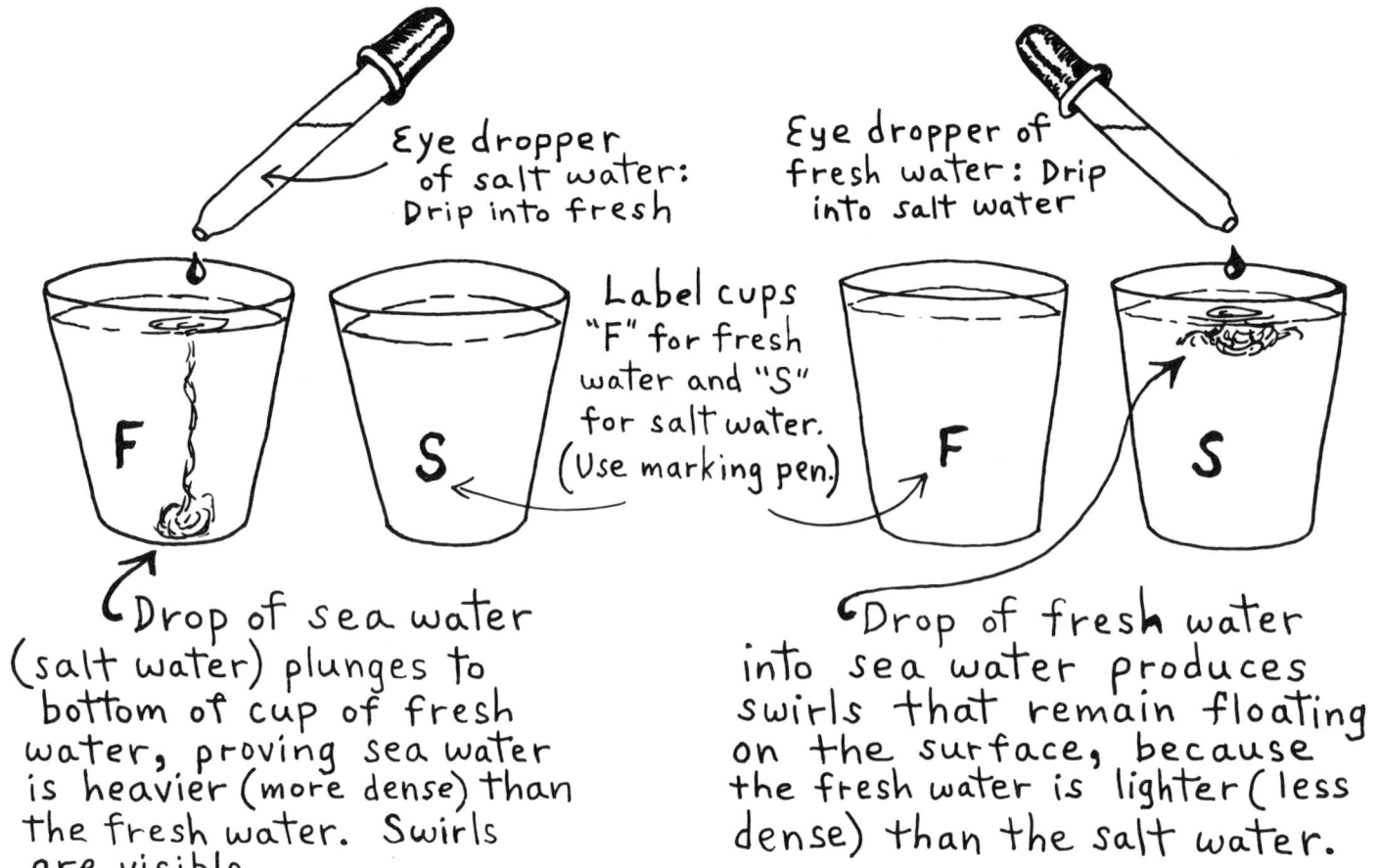

Eye dropper of salt water: Drip into fresh

Eye dropper of fresh water: Drip into salt water

Label cups "F" for fresh water and "S" for salt water. (Use marking pen.)

Drop of sea water (salt water) plunges to bottom of cup of fresh water, proving sea water is heavier (more dense) than the fresh water. Swirls are visible.

Drop of fresh water into sea water produces swirls that remain floating on the surface, because the fresh water is lighter (less dense) than the salt water.

Now, take the other two glasses, one of fresh water and the other of salt water. Set them side by side. Rinse out your dropper under the faucet and fill it with fresh water from the glass. We are going to reverse the experiment. Can you predict what you will see when a drop of fresh water lands in the salt water? Try it! Were you correct? Try it twice more and observe closely.

This time the drop does not fall to the bottom. Instead, it just penetrates the surface and seems to bounce back to the top. Fresh water is lighter, or less dense, than salt water causing it to "float" on top of the heavier salt water.

When a large rainfall comes down on the bay or ocean, what will the effect be upon the water? This is another way in which our marine environment undergoes constant change.

A WHELK'S TINY TREASURES

One of the more common prizes available to curious beachwalkers is a tough, twisted chain of parchment-like discs. When shaken in one's hand, the discs produce a gentle, scratchy rattling.

What have you found? A string of egg cases produced by a carnivorous (meat-eating) marine snail, the whelk.

Each disc on the chain is a separate egg case containing about two dozen miniature snails, each a tiny replica of its parent. By carefully snipping open the disc with a scissor, you can spill out the snails, sometimes only a sixteenth of an inch long, and study them with a hand lens. The discs can also be torn apart with your fingertips, but it must be done carefully. The tiny shells inside are thin and fragile and can be reduced to dust with only slight pressure.

Three large whelks inhabit the Atlantic coastal waters. These are the knobbed whelk (*Busycon caricum*), the channeled whelk (*Busycon canaliculatum*), and the waved whelk (*Buccinum undatum*).

Each of these snails produces a different, though similar, disc-shaped egg case.

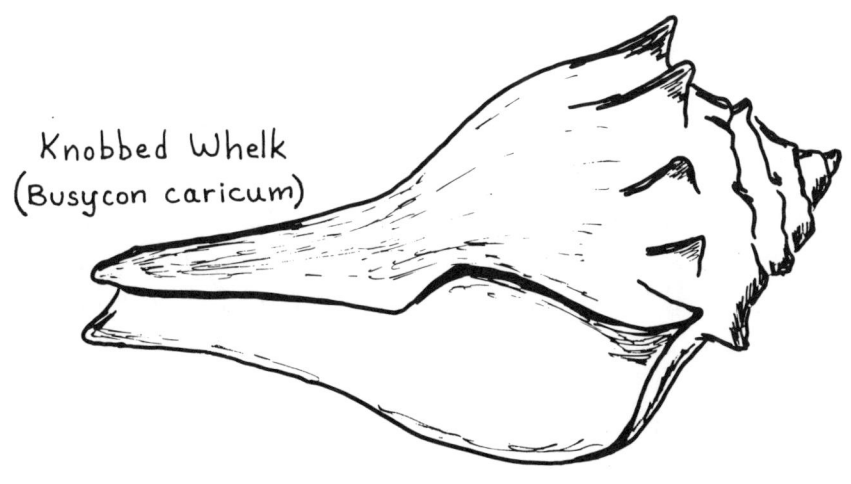

Knobbed Whelk
(*Busycon caricum*)

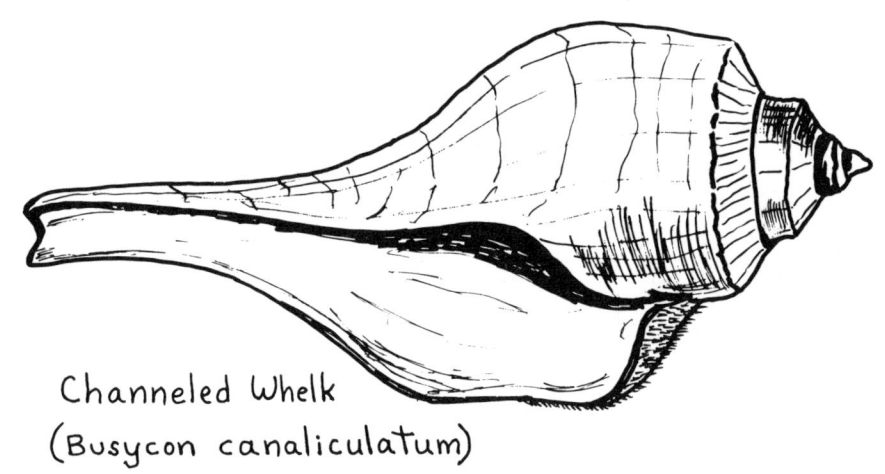

Channeled Whelk
(Busycon canaliculatum)

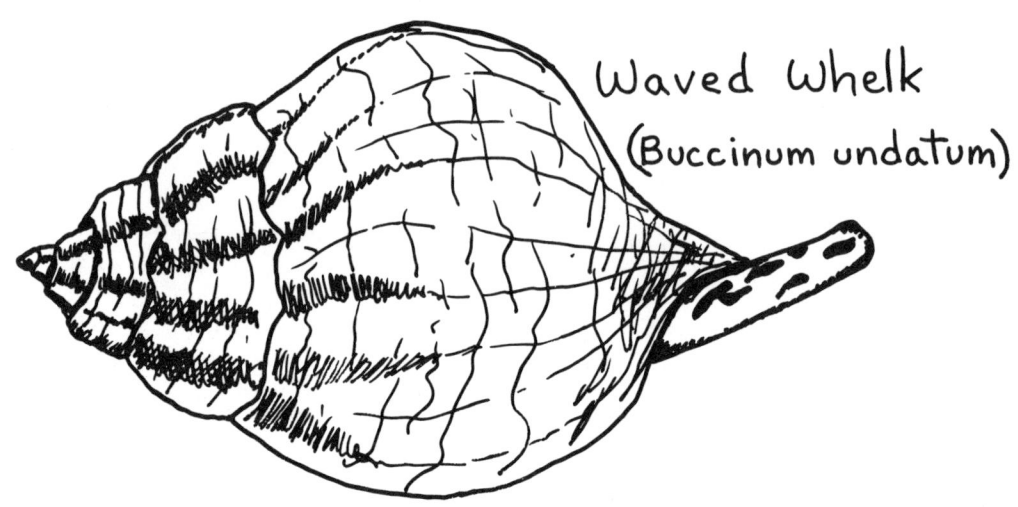

Waved Whelk
(Buccinum undatum)

If you discover the complete shell of an adult whelk, identify it and compare it with its tiny offspring. Are the pronounced "knobs" of the knobbed whelk visible on the babies, or will they develop later?

The egg case of the waved whelk is impossible to mistake. Although the egg cases are disc-shaped, they are smaller (so are the adult snails) than the other large whelks. And the discs are paler in color, rather translucent, and are not laid in chains. Instead, they form a tight cluster in the form of an irregular ball. These are sometimes nicknamed "sea wash balls" because they will produce a foamy lather when rubbed briskly with wet hands.

All of these snails are predators. They hunt for other shellfish, usually mollusks like clams, by plowing slowly along, several inches beneath the sand, with a powerful oversized foot. When the whelk encounters a clam, it grasps the shell with its foot and begins to bore a neat circular hole through the shell with an amazing organ called a radula. This specialized tongue contains rows of rasping teeth. With the clam's shell penetrated, the whelk inserts its proboscis and scrapes away at the soft meat inside.

Both the knobbed whelk and channeled whelk lay their egg cases in long strands up to three feet in length. This, in itself, is near miraculous when you realize that the animals responsible are only seven to nine inches long as adults. The eggs are laid beneath the sand and pushed up as the whelk plows along on its huge foot.

A close look at the discs will reveal whether the parent was the knobbed whelk or channeled whelk. Simply, the egg case of the knobbed whelk has a clearly-defined vertical wall separating the upper and lower surfaces of the disc. The egg case of the channeled whelk lacks this wall, and the discs are fan-shaped with lines radiating from a point on the back of the disc outward to the edges like the spokes of a wheel. Also, the upper and lower surfaces meet to form a single edge with one another. (See illustrations.)

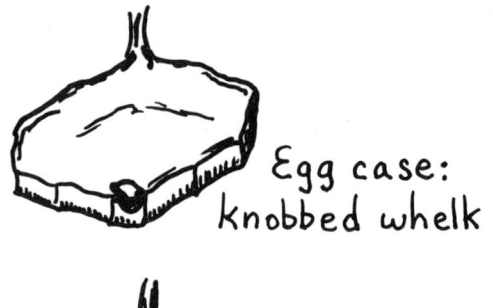
Egg case:
knobbed whelk

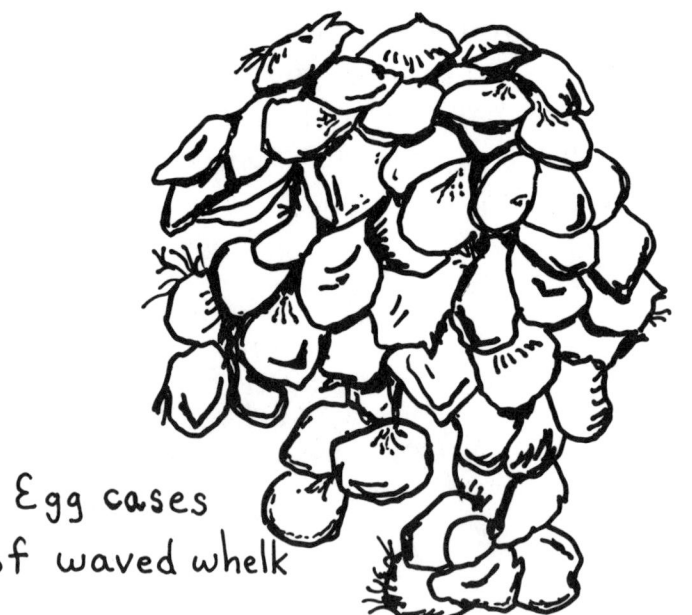
Egg cases
of waved whelk

Egg case:
channeled whelk

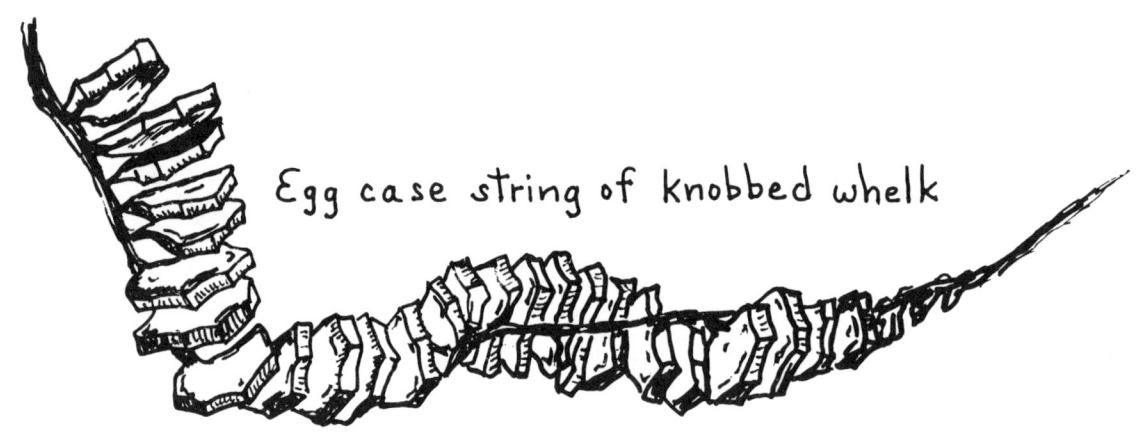
Egg case string of knobbed whelk

If you're having a difficult time locating an adult whelk, ask a lobsterman, as these snails occasionally enter their traps, or try a local commercial fishing boat that uses some form of bottom fishing like a drag net. These boats are usually called "draggers."

Whelks are edible, too. See if you can find a recipe.

The whelk was also important to Atlantic coastal Indian tribes. The shell material was used to make a form of wampum. And the operculum was polished and used in making jewelry. The operculum is a tough plate which seals the opening of the whelk's shell after it has withdrawn inside for protection.

Try to find all three egg cases (the waved whelk may be the most difficult, as it is a deep-water dweller) and mount them on a board with corresponding adult whelk shells.

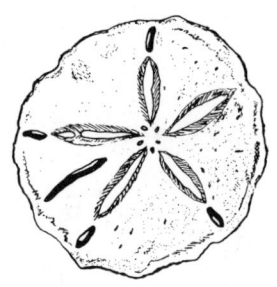

NATURAL WHELK INDIAN RATTLES

The Native Americans of our Atlantic coast were masters at living from their environment. The sea provided them with food, of course, but also with the raw materials for many of their tools of everyday life.

Certain shells were used as scrapers to clean and prepare animal hides for clothing. Others became jewelry, ornaments, or wampum. But how about a musical instrument like the rattle?

The seashore will provide you with the basic materials you'll need to construct an authentic primitive instrument, the whelk rattle.

On your next beach walk, look for a string of whelk egg cases. They are found often on many beaches. The tiny shells inside the dry, disc-shaped egg cases will produce a rattling sound when shaken. (See the preceding unit in this book for detailed information on these egg cases.)

You'll also need a small branch of driftwood, ten to twelve inches in length, for your rattle stick and a few assorted feathers. Gull feathers are a common find at the shore. Small to medium feathers are best. The large flight feathers may look clumsy and overpowering, but try them if you wish. They can always be cut to suitable size.

From the chain of whelk egg cases, cut a section about eight inches long.

Lash one end of the egg cases securely to the top of the stick with several tight wrappings of strong cord or kite string. Knot the string and trim off the ends.

Bend the egg case chain into a gentle arc and lash the other end securely to the middle of the stick, as shown in the illustration. The leathery, parchment-like material of the egg cases is quite strong. It can be bent and twisted without breaking.

41

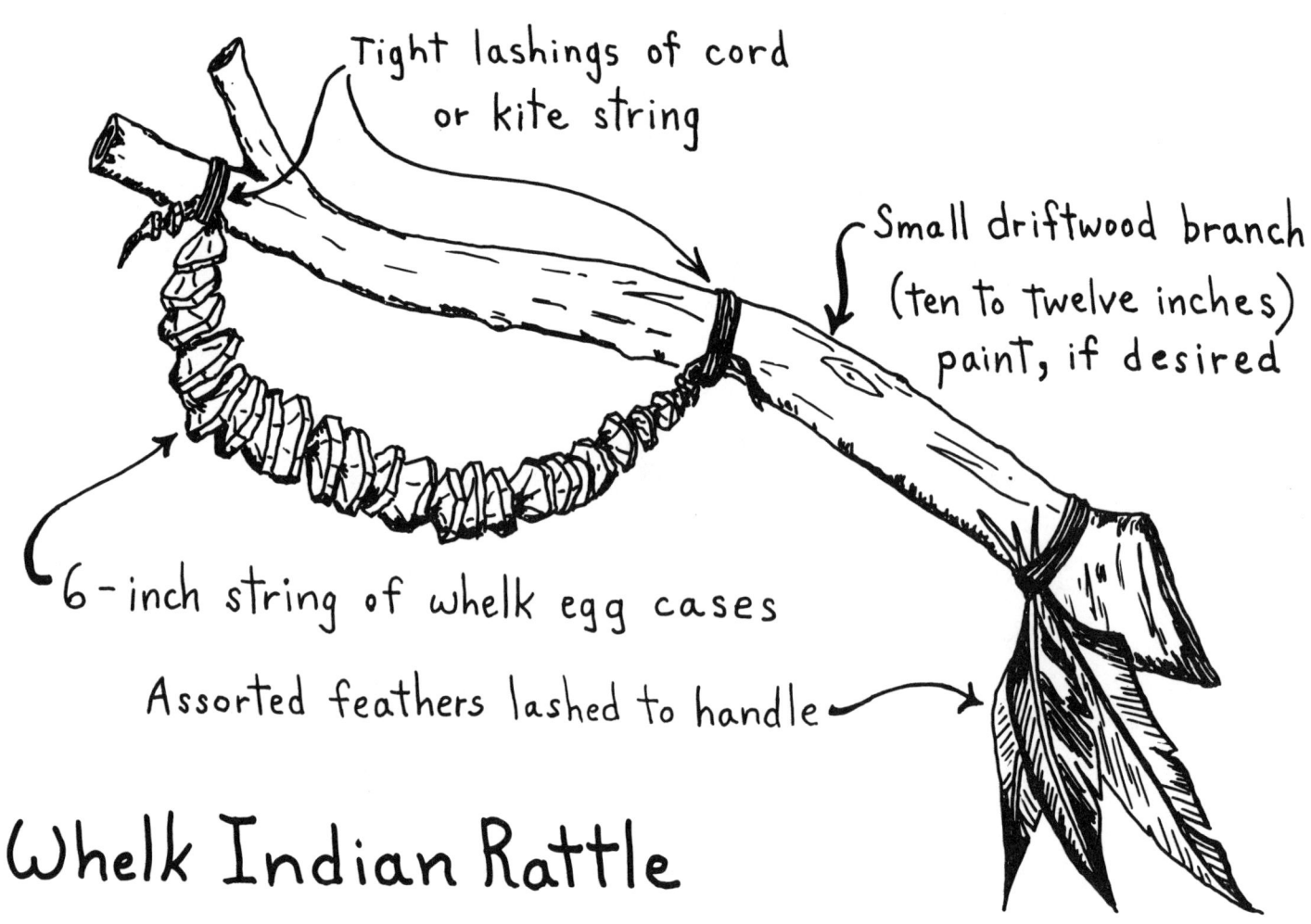

Tight lashings of cord
or kite string

Small driftwood branch
(ten to twelve inches)
paint, if desired

6-inch string of whelk egg cases

Assorted feathers lashed to handle

Whelk Indian Rattle

Now, add your feathers to the base of the handle, securing them tightly with neat lashings of string.

Your driftwood rattle stick may be left in a natural state of weathered gray, or it may be carefully painted with American Indian designs. The decision is yours. But don't rattle your brain thinking about it.

SUNDIALS ON THE SHORE

We often use the term "clockwise" to describe the direction in which a motion is made, or to describe how something rotates, or turns. But why did the inventors of early mechanical clocks decide that the hands should rotate from the upward position at twelve o'clock to the right?

Building this simple sundial may give you a clue. You'll also learn a bit about the earth's motion and its relative position to the sun by examining the length of shadows cast on your sundial and the distance between these shadows from hour to hour.

On your next trip to the beach, bring these materials:
- A wood dowel, ¼-inch in diameter and 18 inches long
- A sheet of oaktag (30" by 24" is standard size)
- A simple trail compass
- A ruler and a marking pen

Along the center of each edge of the oaktag, write the letters, N, E, S and W to represent the four cardinal points of the compass, as shown in the illustration.

Lay the oaktag down on a flat, clear area of sand away from the water.

Use the trail compass to find the direction of magnetic north and turn your oaktag so that the "N" on the edge faces north as indicated by the compass.

Place stones at each corner of your paper so it will not be disturbed by the wind.

The dowel will be your shadow stick. It must be pushed several inches into the sand through the center of the oaktag sheet. To find the exact center of the oaktag, draw two straight lines connecting each pair of opposite corners. These lines are called diagonals. Use your ruler or the edge of your dowel to draw the lines straight. The point where these lines intersect, or cross, is the center of your paper. Push the dowel firmly through this point into the sand beneath. Be sure the dowel is vertical. With the sun shining, a shadow will be cast across the paper.

Lay your ruler along the center of this shadow and draw a line with the marking pen. Draw the line exactly as long as the shadow from the base of the dowel outward.

Look at your watch and record the actual time by writing it along the shadow as shown in the illustration.

Sundials on the Shore

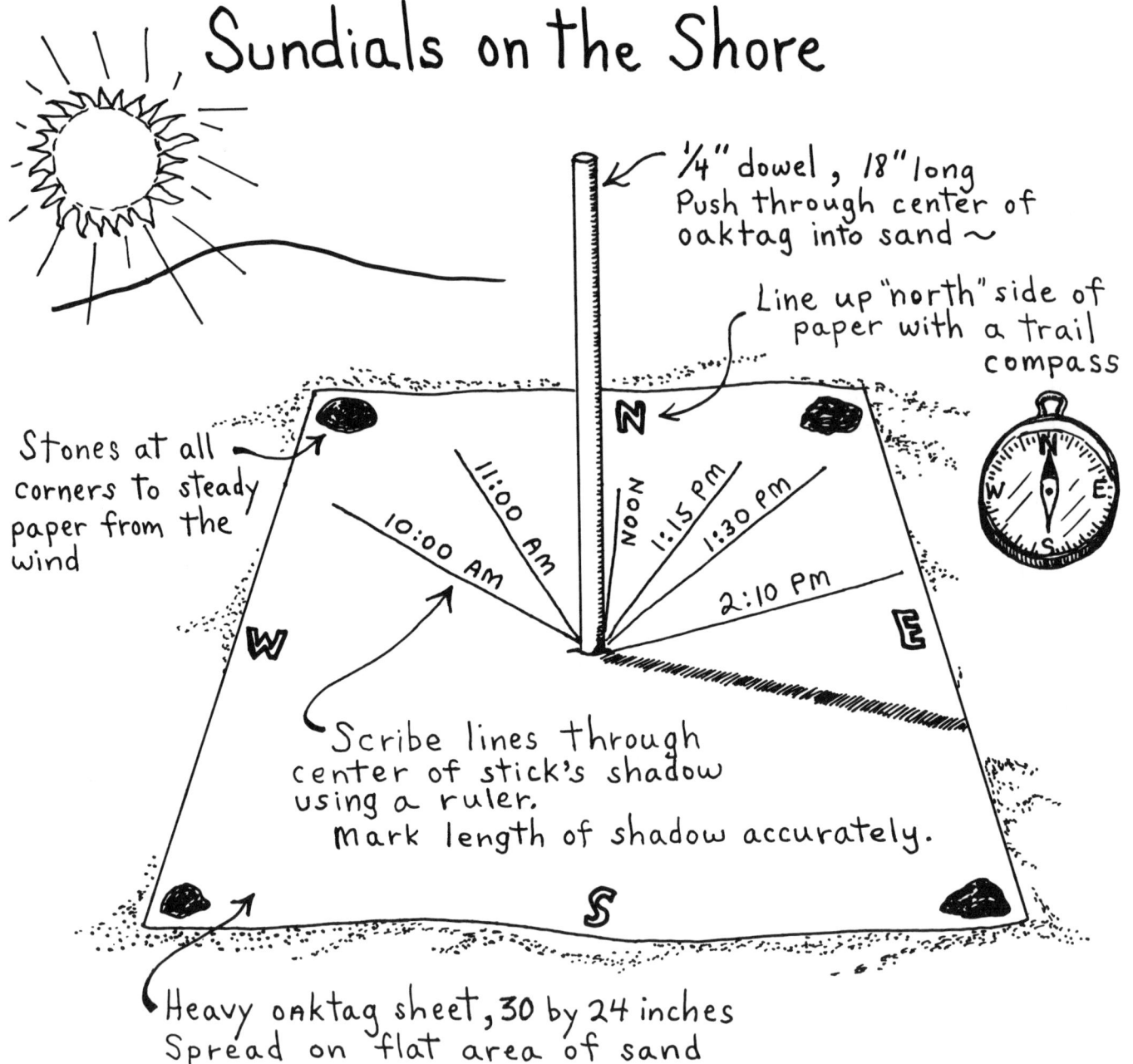

¼" dowel, 18" long
Push through center of oaktag into sand ~

Line up "north" side of paper with a trail compass

Stones at all corners to steady paper from the wind

11:00 AM
10:00 AM
NOON
1:15 PM
1:30 PM
2:10 PM

N

W

E

S

Scribe lines through center of stick's shadow using a ruler.
Mark length of shadow accurately.

Heavy oaktag sheet, 30 by 24 inches
Spread on flat area of sand

- In which direction is the shadow moving?
- At what time is the shadow the shortest?
- In what direction does the noon shadow point?

Continue to record shadows and times in this way at various intervals during your stay at the beach. (Let's hope the sun keeps shining.)

Now, let's examine your work.

Does your shadow travel in a "clockwise" direction?

Sundials were in use long before mechanical clocks, and people were already accustomed to the motion of the shadows on their sundials. So, when mechanical clocks became a reality, the hands were designed to turn in the same way as the movements of the shadows cast by the sun.

Look also at the lengths of the shadows. At what time was the shortest shadow cast?

Consider these questions also:

In what direction does the noon shadow point?

Are the hourly shadows all the same distance apart?

Do you notice any interesting patterns?

The angle of the earth to the sun changes throughout our seasons. Try this activity in summer, fall, winter and spring. Save the sundials and compare them. What remains the same? What changes do you notice in the shadows you have recorded from one season to the next?

Let your sundial be a "shining" example of your creativity and scientific ability.

TELLING TIME ON YOUR FINGERS

This activity is a simple survival skill for campers and hikers, but it may come in handy at the beach, too.

People who enjoy recreational camping in remote wilderness areas often leave their campsites for a day hike. But it's important to be back before nightfall. It is not uncommon for careless campers to get lost at night in the woods.

Perhaps you've taken a trip to the beach and promised to be back before dark. But you forgot your watch! And you don't know how soon the sun will set.

In either case mentioned above, you don't have to worry if you have your fingers...because that's all you'll need to determine how much time remains before sunset.

Simply face the setting sun and fully extend your arms out in front of you. Bend your wrists so that your palms face inward as in the illustration.

How many fingers can be placed in the space between the bottom edge of the sun and the line of the horizon? (Do not use or count your thumbs.)

Each finger-width is equal to fifteen minutes of time. If you can fit two fingers between the sun and the horizon, the sunset will take place in thirty minutes (2 x 15), or a half-hour.

Of course, your fingers won't tell you the exact time of day the way a watch would, but they will tell you when to expect darkness.

Most beaches work well for this activity because clear, unobstructed horizon lines are common. The sun usually sets on the open landscapes of dunes or on the flat expanse of sea. Try it to see how accurate your estimations are. You might even try counting a fraction of a finger for greater accuracy. A third of a finger-width would equal five minutes.

Telling Time on Your Fingers

Each "finger space" between the sun and the horizon represents fifteen minutes.

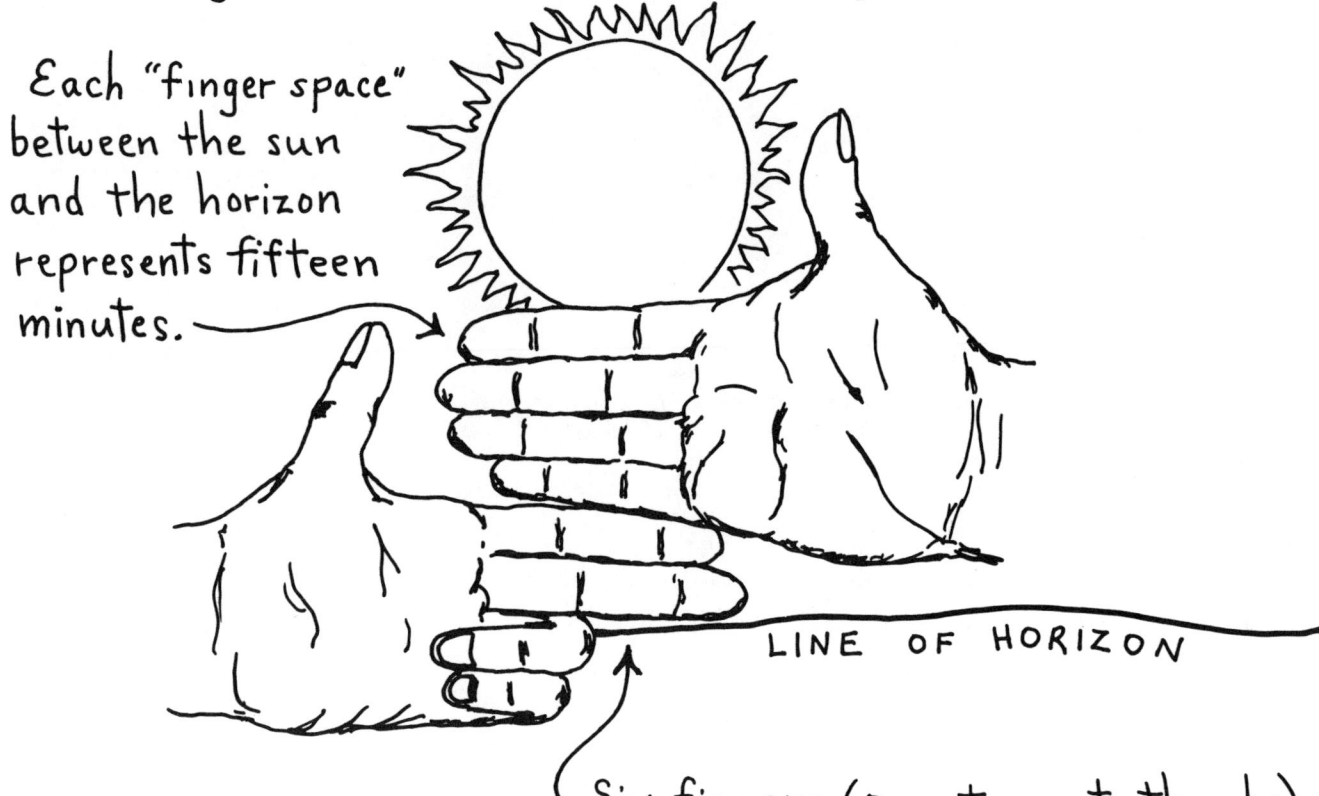

LINE OF HORIZON

Six fingers (Do not count thumbs) between sun and horizon means that sunset will occur in 90 minutes (6x15), or in an hour and a half.

My fingers tell me it's getting late! Let's move on to another activity.

WATER SCOPES

Did you ever stand at the water's edge of a quiet bay and stare into the water? What could you see beneath the surface? Probably, not very much. At best, you might see dark, patchy areas that you guessed were beds of eel grass or algae.

Chances are that many small fish, perhaps entire schools, were darting about unseen by your eyes. Or, maybe some crabs were skulking and clambering along the bottom.

The water's surface acts like a mirror, reflecting the sunlight back at your eyes. Many of the sun's rays are scattered and bent by the water. This further reduces your ability to see clearly.

A simple device, the water scope, can reveal clearly the world of life in the shallows beneath you. Let's make one!

Water scopes may be made in many sizes; a small one from a 10-ounce soup can, a medium sized one from a 1-pound or 2-pound coffee can, and a large one from an old metal or plastic pail. The construction method for all three is the same.

Choose your container and remove the bottom. Both top and bottom ends should be open.

Cut a sheet of heavy-duty, transparent plastic wrap, (the kind used for food storage), large enough to cover one end of your container. Be sure the plastic is big enough to overlap the sides of your can or bucket.

Begin sealing the plastic wrap around the sides of the can with several tight wrappings of heavy cord. (You can hold the plastic wrap temporarily in place with a rubber band.)

It is important that your water scope be completely watertight. To insure this, wrap several turns of 2-inch cloth duct tape around the edges of the plastic wrap where it meets the sides of the can. Duct tape is available at any hardware store. Compare your finished water scope with the illustration.

Water Scopes

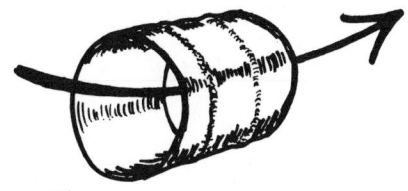

1 lb. or 2 lb. coffee can, or large juice can.
Remove both ends.

One end sealed with heavy duty clear plastic food wrap.
Secured with several wrappings of strong cord and cloth duct tape to prevent leaks.
Wrap Tightly!

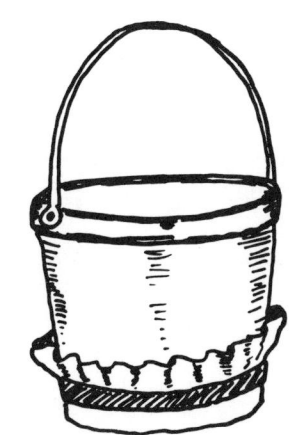

Large water scope made from plastic pail.
Follow procedures as detailed above.

Now, go out and try it! Wade out, waist deep, into the water. Remain still for a few minutes, so as not to disturb any fish. Push the sealed end of the water scope slightly below the surface. What do you see?

Do you notice that the water scope also magnifies what you see? This happens because water is pressing against the plastic wrap, causing it to bulge inward and form a curved surface. This acts like a *convex* lens to magnify the images you see. Do not allow your water scope to fill with water, as it will not work.

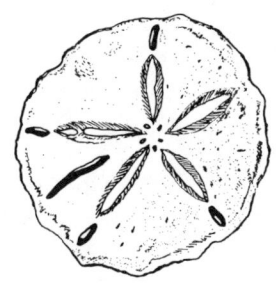

BURLAP BEACH WEAVINGS

Would you like to create an interesting and accurate record of your next trip to the beach? A record that is also an art form? This is it! A burlap beach weaving.

Burlap can be purchased readily at many home garden centers, hardware stores and tree nurseries. Even a large potato bag will do nicely. Just snip open the side and bottom seams. Try to get a piece at least three feet by three feet in size.

Tie a sturdy branch or length of driftwood to the top and bottom edges of your burlap. You may work on your weaving on a flat surface, or while it is suspended in air by hanging the upper branch with lengths of cord.

Your weaving materials will be collected on your next beach walk. Pick up anything which can be passed over and under the *warp* and *weft* of the burlap material. The warp refers to the vertical threads of the fabric. The weft refers to the horizontal threads that are woven into the warp.

The materials you collect may be woven vertically or horizontally. Burlap has a wide, open weave. The threads can be stretched and separated easily to allow larger items to be woven into the existing fabric. You can even snip a thread here and there to get wide items into your weaving.

When collecting natural seashore materials, be sure items like shells or crab carcasses are clean of any small bits of meat. If they are not, discard them or boil them clean. Algae, or "seaweed," may be woven while wet and allowed to dry on the weaving.

When your weaving is complete, use it as a decorative wall hanging.

Burlap Beach Weavings

Tie burlap to lengths of driftwood, top and bottom, or to dowels. Insert screw eyes into upper pole and suspend your weaving from a cord for display. Burlap threads separate easily for weaving purposes. Try almost anything!

What Can I Weave?

The following items may be used successfully in your burlap beach weaving. Can you think of others?

crab claws	dried algae
sea shells	whelk egg cases
grasses	flat stones
feathers	animal bones
driftwood	pieces of rope
old fishing line	skate egg cases

litter (flattened cans, paper, plastic wraps, etc.)

Remember! You are creating art, but also a record of the beach environment. Be free in your choices of collectible items. Almost anything goes.

TIDE STICKING

Time and tide wait for no man. Have you heard that saying? We simply can't prevent the passing of time, nor can we control the incoming flow of ocean waters we call the tide.

Tide and time are closely related. We have tide tables and calendars that tell us when tides are high and when they will be low. The sailing schedules of large ships and sailing vessels often depend on the tide calendar.

The tides are caused by the attractive gravitational forces of the sun and the moon upon the earth. A complete rise and fall of the surface of the oceans and seas occurs twice in every 24 hours and 50 minutes. (This is a lunar, or moon, day.) Because of this time difference compared to the earth day of 24 hours, the tides change at a slightly later time each day. Rivers are also affected by the tides, especially near their mouth, where they flow into the sea.

Here's a simple activity that will allow you to observe and measure the change in tides over a short period of time. You'll need only a meter stick and tape measure, and the activity sheet provided.

If the wave action is heavy at the ocean, try this activity in a quiet bay or cove.

Pick a smooth, clear area of beach and push the meter stick into the sand at the water's edge. Push the meter stick in deeply enough to remain firm if the tide is coming in. Twenty-five or thirty centimeters should be sufficient. (See the illustration.)

Every ten minutes, make a measurement with your tape from the base of the stick to the changing water's edge. Chart these numbers on the graph provided. Take readings for one hour or longer.

Is the tide coming in or is it going out? Does the tide travel the same distance each ten minutes?

TIDE STICKING

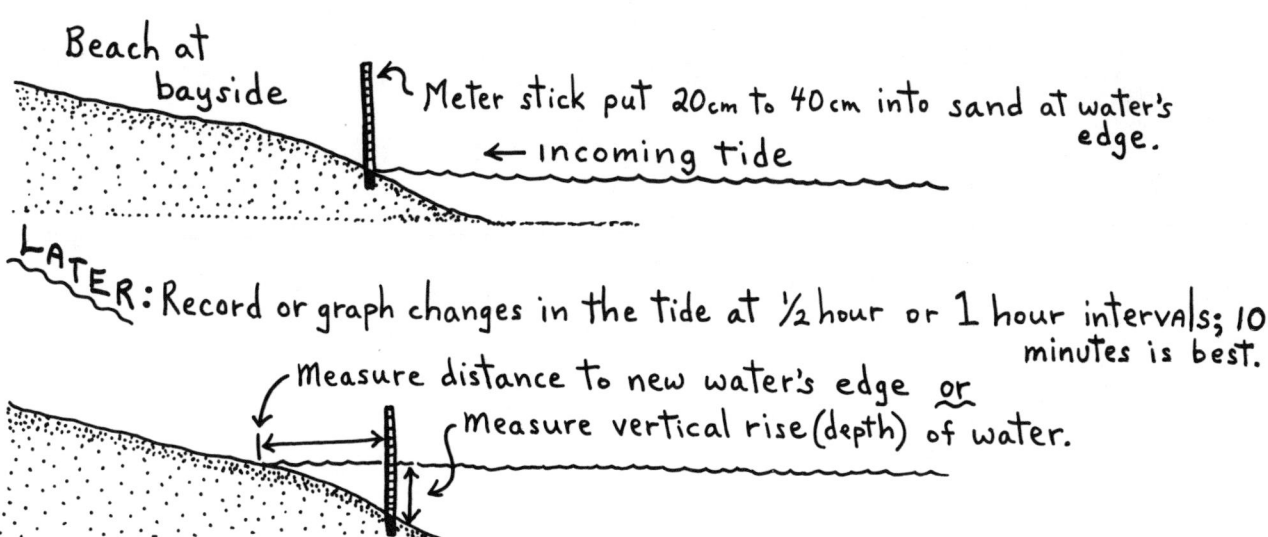

Beach at bayside — Meter stick put 20cm to 40cm into sand at water's edge.

← Incoming tide

LATER: Record or graph changes in the tide at ½ hour or 1 hour intervals; 10 minutes is best.

Measure distance to new water's edge or measure vertical rise (depth) of water.

If the tide is out-going, your stick will remain high and dry, but if you have an in-coming tide you can also take vertical measurements of the water's changing depth by reading the meter stick. How high has the water risen in ten minutes? Remember to subtract the length of stick buried in the sand. For example, if your stick was buried to the 25 cm mark when you started, your actual water height would be zero. If, in ten minutes, the water level is at 37 cm, your change in water height is really 37 cm – 25 cm, or 12 cm. A chart is provided for you to record this information. Much of good scientific work depends upon the accurate recording of data, or information.

Is the water level rising at the same rate every ten minutes? Try to construct your own graph to show this new information.

If you're spending several hours at the beach, perhaps fishing, swimming, or sunbathing, you can record the changes over a longer period of time.

Certain places have unusually fast-flowing, powerful tides that can be dangerous. Work at a beach with which you are familiar.

By the way, if you allow the water to rise too high on your meter stick, it may be lifted from the soft sand by the water and drift away. Keep an eye on it.

You can learn more about the importance of tides in our environment, how fishing is affected by them, and about special kinds of tides by using an encyclopedia. Do you know what a *neap* tide is? And a *flood* tide? Or an *ebb* tide? How about a *spring* tide? Don't be fooled by the names. They may not mean what you think they do. Can a strong windstorm affect the tidal flow?

Find out! And good tidings to you in your work!

TIDE GRAPH for CHANGE in WATER'S EDGE ～

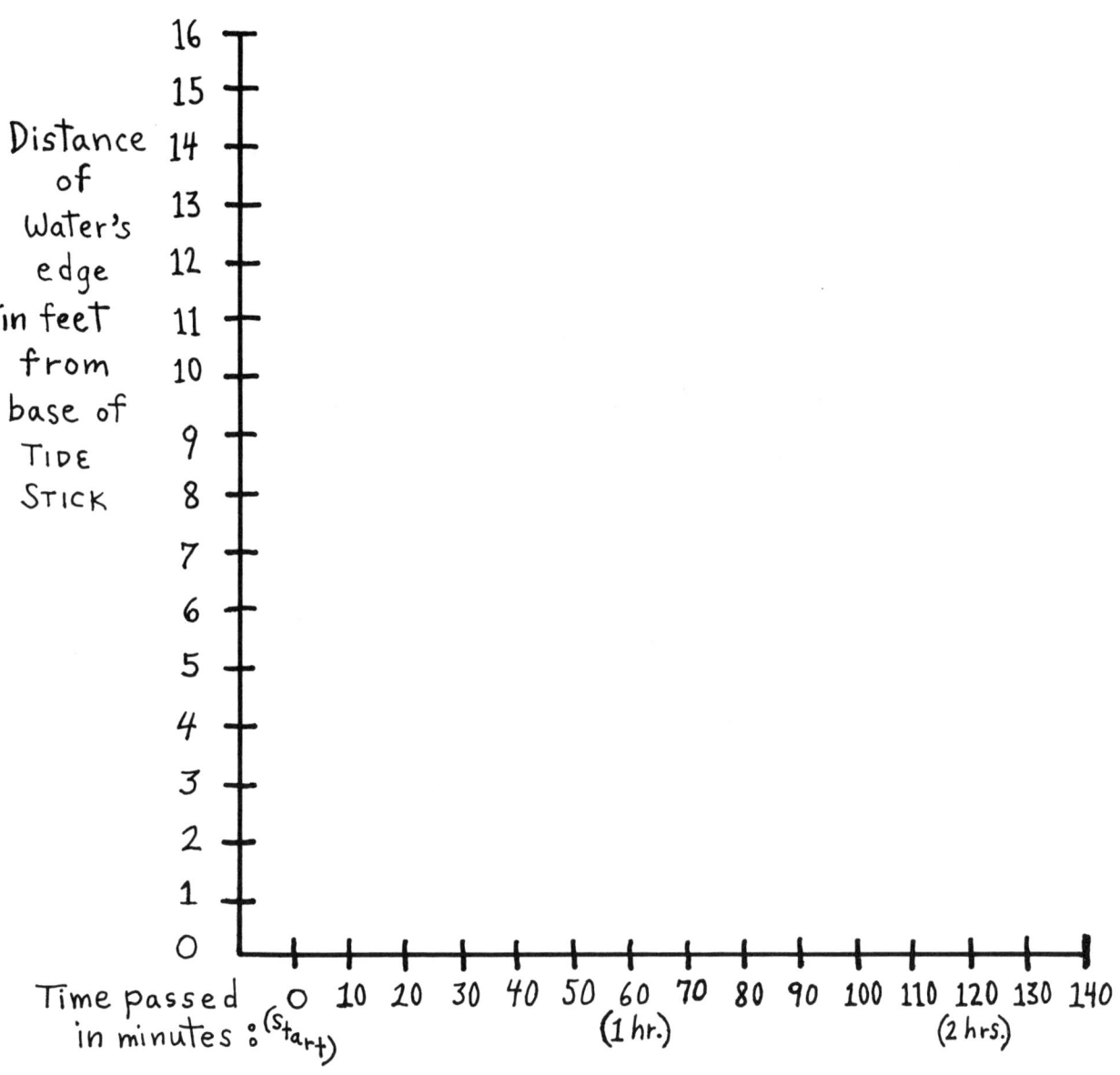

Distance of Water's edge in feet from base of TIDE STICK

16 —
15 —
14 —
13 —
12 —
11 —
10 —
9 —
8 —
7 —
6 —
5 —
4 —
3 —
2 —
1 —
0

Time passed in minutes : 0 (Start) 10 20 30 40 50 60 (1 hr.) 70 80 90 100 110 120 (2 hrs.) 130 140

* Is your tide incoming or outgoing? _____

Chart for Change in Water Level (Incoming Tide only)

Time Passed	0 min (start)	10 min.	20 min.	30 min.	40 min.	50 min.	60 min (1 hr.)	70 min.	80 min.	90 min.
Height of water in centimeters	0 cm									

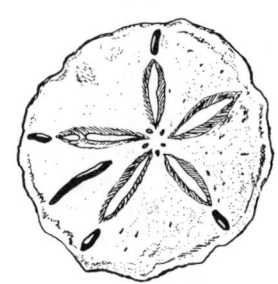

INDIAN PIT COOKING

The next time you're planning to spend a full day (or evening) at the beach with family or friends, why not try a Native American cookout?

Pit cooking is still practiced today, even in the South Pacific islands, to prepare seafood for feasts or just as an old-fashioned "clam bake."

Your oven is a pit dug out of sand on the beach. The size of your pit depends upon the amount of food being prepared. But even for a small group, make the pit at least eighteen inches deep.

Collect dry grasses and firewood from the area. Your firewood should include small kindling pieces, like twigs, to large and medium-thick branches or driftwood. Dried grass or paper can be placed at the bottom to start the fire.

Build up a layer of smaller wood pieces at the pit bottom and use larger pieces at the top.

Place fist-sized beach stones at random within your heap of wood, including a final layer of stones across the top of your firewood heap after it has been set aflame. See the illustration.

Light your fire and allow it to burn down to hot coals and ash. The super-heated stones will fall to the bottom as the fire burns down.

Begin laying your foods into the pit between layers of wet seaweed. Clams, mussels, oysters, crabs, shrimp, corn and smaller fish can be steamed to juicy perfection in your pit.

Try wrapping corn and fish in foil. Shellfish can be placed directly into the pit.

Work quickly while the stones are hot. The wet seaweed and natural juices of the shellfish will provide some moisture to produce steam, but add a cup or two of sea water just before sealing up the pit with a heavy layer of beach grasses or leaves. Have these at hand to be placed over the pit immediately to prevent the loss of heat and steam.

Shellfish like clams and mussels will open when cooked and are ready to eat as they are, or can be dipped in melted butter.

Indian Pit Cooking

Step 1.

Line bed of deep fire pit with stones.

←Medium large stones mixed with wood. Build a roaring fire.

Dried grass and small twigs used as kindling.

Step 2.

Build up alternate layers of seaweed (kelp, rockweed, sea lettuce) with clams, oysters, mussels and corn.

Seal top of pit with heavy layer of grasses or reeds.

Super-heated stones settle at bottom as fire burns down. Pour water on them to produce steam.

If your fire was roaring hot, your food may be ready in as little as ten minutes! When your activity is over, be sure to extinguish your fire with water and fill in the pit with sand.

Not only is this a great way to add interest to your beach party, but you'll gain a better understanding of everyday life in an early, coastal Native American tribe. Enjoy!

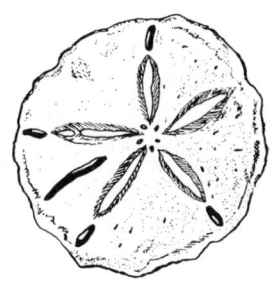

SAND CASTING

Beach castings with plaster of Paris are fun and easy to create. They produce a durable art form that can be hung or displayed on a shelf. There are several methods to be discussed, providing a variety of artistic casts, but all require a plaster of Paris mixture. Plaster of Paris can be purchased in twenty-five pound bags at any hardware store for a very reasonable price.

In all casting work, never mix the plaster until you are ready to pour it onto your prepared artwork. Plaster sets quickly and using salt water in your mix at the beach actually speeds up the process. Mix the plaster according to directions on the bag. If you are working with a group at the beach, buy at least two bags.

You'll need a sturdy stick to stir the plaster mixture and two plastic pails, one for mixing and one to hold water. Rinse the mixing bucket after each cast.

Your plaster should not be too thick. It should flow from the pail like melted ice cream.

Allow the plaster to set undisturbed in the sun for a half hour or more. Carefully brush the sand away from the edges and lift out the cast. Your cast may feel quite warm, as plaster gives off heat as it sets.

Loose sand can be brushed off with an old paintbrush and any rough edges can be smoothed with coarse sandpaper. The plaster is still rather soft and easy to work with at this point. It will take a few hours for the plaster to thoroughly set and harden.

If you used shells or stones in your cast, they can be highlighted by hand painting them with varnish or shellac.

Now, let's discuss the various techniques you may wish to try.

Sand Casting!

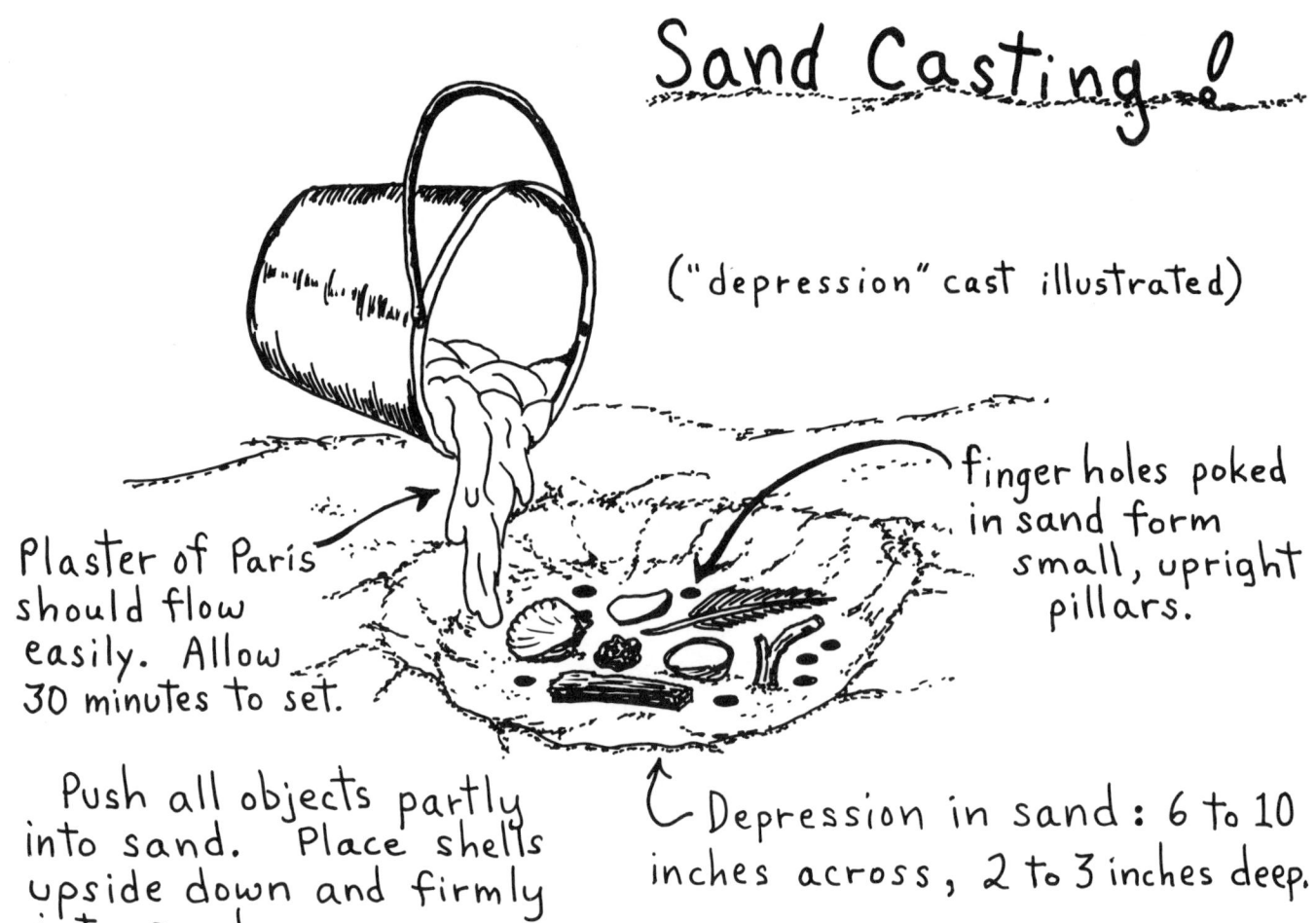

("depression" cast illustrated)

Plaster of Paris should flow easily. Allow 30 minutes to set.

finger holes poked in sand form small, upright pillars.

Push all objects partly into sand. Place shells upside down and firmly into sand.

Depression in sand: 6 to 10 inches across, 2 to 3 inches deep.

Depression Casts: This is the simplest form of casting and still produces nice results. As its name suggests, you scoop out a small bowl-shaped hollow or depression in the beach sand. Always brush away the dry surface sand and work in the damp sand beneath, as your depression will better hold its form. Keep the depression small, about 6 to 10 inches across and 2 to 3 inches deep. Otherwise, you may use more plaster than needed and your cast will be excessively heavy. The simple casts of one large shell or a small arrangement of items look best.

Collect the materials for your cast. These may include driftwood, shells, stones, pebbles, or feathers. Arrange them attractively in your sand depression.

You must think "upside-down" in casting because what faces down in the sand will face up on the cast. If you want the colorful outer surfaces of shells to show, place the outer surface down in the and. Press all your casting objects at least halfway down into the sand. You can add miniature stone columns, resembling the stalagmites of a cavern, by poking holes into the sand with your fingertip.

Mix the plaster to a creamy consistency and pour it into the hole, covering all your objects. Smooth the surface of the plaster quickly by hand or with a flat stick and allow it to set. Remove it carefully and brush away any loose sand.

Frame Casting: These casts can be used as hanging plaques. Instead of a depression, smooth the sand surface flat and level. Place a small wooden frame, two inches high, constructed of scrap wood, onto the sand.

Arrange your objects inside the frame, not too close to the edges, and press them slightly into the sand.

Pour in your plaster mix. Place one or two jumbo paper clips into the wet plaster near the top of the cast to serve as hangers. Or, make a hanger of thin rope by taking a six-inch length and knotting both ends. Push each knot well into the plaster near opposite sides of the frame. The knots must be fully buried in the plaster, but don't push so hard that you will disturb your artwork on the underside.

When the cast has fully set, gently lift out the frame and carefully pop out your casting. If the wood frame is securely set in the plaster, you may wish to leave it. The frame may be painted.

Fish and Shell Fossil Casts: Imitation fossils can be created in depression casts or frame casts. These are impression casts. The objects cast in plaster will be removed leaving only their impressions.

With a small, stiff paintbrush, spread a thin layer of petroleum jelly over the top surface of a shell or fish. (Scallops and flounders work beautifully.)

Place the object on the sand, painted side up, in your depression or frame. If using a fish, press it lightly into the sand. Only the top half should show with no gaps between the fish and the sand. You don't want plaster to flow beneath it, making it difficult to remove later.

Mix the plaster and pour it. Let it set and lift out the cast. The petroleum jelly prevents the fish or shell from sticking to the plaster, making it easy to remove and leaving an accurate impression.

When the cast is fully dry, wipe out or wash out any excess petroleum jelly. The cast may be painted if you wish.

You may find that petroleum jelly coatings are unnecessary on fish because of the natural coating of mucus found on the skin of fish. Even shells with a smooth surface can be popped out of the plaster without being coated with petroleum jelly. See what works best for you.

All the techniques produce exciting results. What creative ideas can you dream up? Try them all.

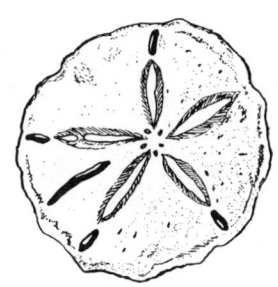

ANIMAL TRACK CASTS

The damp sand of the beach and the soft mud of the marsh provide ideal locales for creating accurate plaster castings of animal tracks. Both environments support an abundance of animal life which includes shore birds, deer, foxes, raccoons and rabbits. If working with a large group, you'll need a twenty-five pound bag of plaster, but if you are working alone in the field, you'll want to keep your materials to a minimum. Let's assume you're alone.

Place two or three pounds of plaster of Paris into a sturdy sealable plastic bag or into a large coffee can with a tight-fitting plastic lid. This is enough plaster for several casts.

A large empty juice can (#10 size) will serve as a mixing can. The top lid should be completely removed. You'll need a stirring stick, a water container and several heavy paper strips of oaktag or construction paper. Cut the strips into one inch (2.5 cm) widths and twelve inch (30 cm) lengths. Bring along some medium to large paper clips, also.

Look for a good clean track, one with a well-defined impression. Do not cast the first track you see unless it is suitable.

Carefully pick out any leaves, pebbles, or grass that may have blown or fallen into the track impression.

Take a paper strip and form it into a collar around the track. A twelve inch strip will form a circle with a diameter, the distance across, of about four inches. For larger tracks, such as deer, leave the circle full-sized. Overlap the two ends and secure them with a paper clip. For smaller tracks, overlap the edges a bit more, but always leave a half-inch of space between the wall of the paper collar and the track.

Press the paper collar gently into the ground around the track to prevent the plaster mix from seeping out.

Making a Plaster Animal Track Cast

First cast is called a "negative."
(track is raised)

Re-cast the negative to make a "positive." (Track is depressed as in natural form.)
Paint, if desired.

Surround track with paper collar (1" x 12"); overlap edges to correct diameter and seal with paper clip. Pour in plaster.

Mix the plaster according to package directions. Generally, this is two parts of plaster to one part of water. If you plan to use salt water from the beach or marsh, remember that your mix will set up more quickly than it would with fresh water and you'll have to work fast.

When the mix is like thickened cream, pour it into the paper collar, being careful not to disturb the track.

After a few minutes, as the plaster begins to set, a large paper clip can be inserted at an angle halfway into the plaster near the edge. This will serve as a hanger later.

Allow the plaster to set for a half-hour to an hour and gently lift up the track cast, paper collar and all. It will take several more hours for the cast to fully harden at home. Only then should you brush away any excess dirt stuck to your cast and peel away the paper collar.

Your first casting is a negative, as the track appears as a raised form on the plaster base. You may paint this and hang it for display or you may use it to make a positive cast. A positive cast shows the track as it appears in nature, as a depression.

To make a positive, wrap a paper strip two inches (5 cm) tall around the negative cast and tape or clip it closed.

Use a small stiff paintbrush to coat the entire surface of the negative with a thin layer of petroleum jelly.

Mix a small batch of plaster and pour it into the collar. Let it set fully and peel away the collar. (If you wish to hang your positive cast, don't forget to insert the paper clip after pouring the plaster.)

Gently push the blade of a butter knife between the casts. They should separate easily due to the petroleum jelly. This can be washed off later when the cast has fully hardened.

If you're unsure as to what animal created the tracks you have cast, there are excellent field guides available on track identification at libraries or bookstores.

As an extension to this idea, try making leaf print casts. They make nice imitation fossils.

DRIFTWOOD COLLAGES

Every beach has its own special character. It may be in the particular color or texture of the sand, in the unique assortment of shells found in that area, or in the type of birds, animals, or plants that thrive there.

The character of the beach in your area, or of any beach you may visit on vacation, can be captured and brought home by creating a driftwood collage. Your completed collage may become a permanent decoration in your room.

You'll need a piece of driftwood. Look for a moderately sized piece of an interesting shape that can stand on its own if placed on a table or shelf. The more grayed and weather-beaten it is, the better.

Collect a wide assortment of beach objects. You may include sea shells, bones, feathers, pebbles, egg cases, dried algae (seaweed), beach grasses and other plants, crab claws and shell casings, and a scoop of beach sand. Late fall and winter are prime times to collect, and the driftwood collage can be a pleasant indoor craft or hobby.

Collect far more items than you'll actually use in your collage. This allows you more choices as you compose your objects on the driftwood by size, shape, texture or color. Try many arrangements.

When satisfied with the composition of your collage, glue each item in place with white glue or household epoxy. Larger items may be taped in place while the glue sets. Remove the tape later.

Thin layers of white glue should be brushed on selected areas of the driftwood. Sprinkle beach sand onto the glue to create textured sandy patches.

The surfaces of your shells may be highlighted by hand painting them with varnish.

If you're lucky enough to travel overseas to beaches of other nations, your collage may serve as a scientific record of that beach, as well as a fine piece of art.

DRIFTWOOD COLLAGE

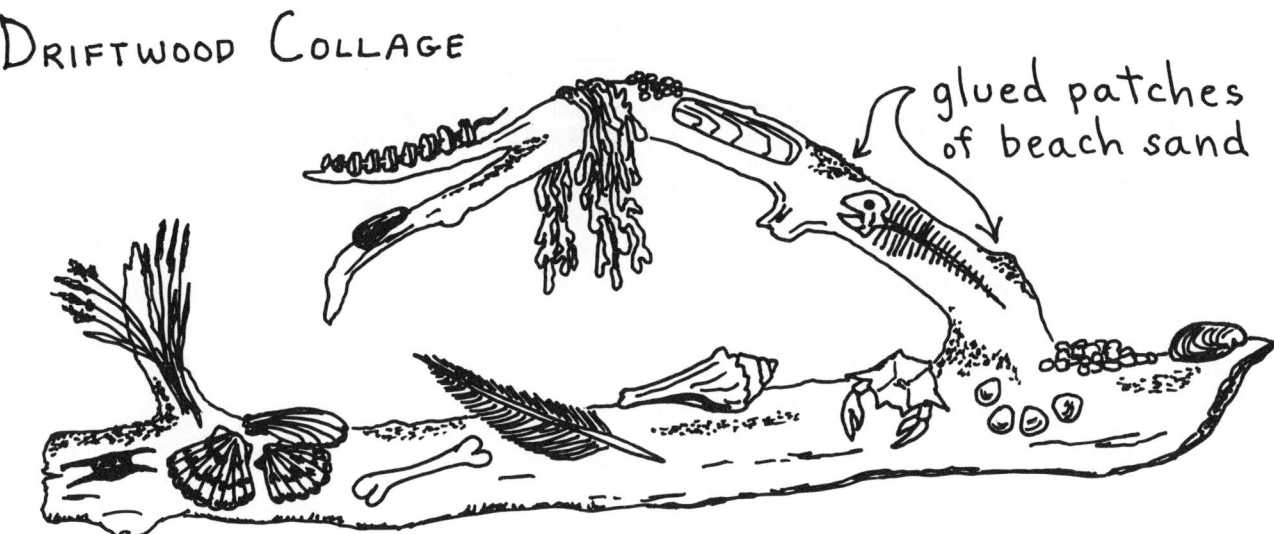

glued patches of beach sand

Secure items with white glue or epoxy. Try: dried algae, fish and animal bones, feathers, whelk and skate egg cases, beach grasses, pebbles, berries, crab parts + assorted shells.

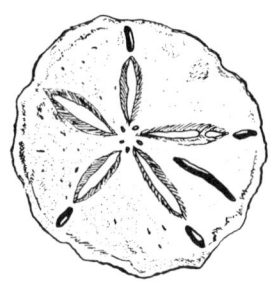

MARINE MOBILES

Mobiles are suspended art forms that display balance and free-swinging motion in their construction. Art works have a theme. Your theme will be life in the sea.

The illustration will provide ideas to guide your assembly. Almost anything found on the beach can be hung on your mobile.

Start with a large main bar, perhaps of driftwood. Use strong thread or plastic fishing line to suspend your seashore treasures and leave all the knots a bit loose so that adjustments in balance may be made as you add items. Simply slide the objects to new positions on the cross bars until they hang rather level.

Add some smaller cross bars as you work down. Be sure all the objects can swing about freely without striking each other.

Algae should be dried before hanging. Attach shells by tapping a tiny hole with nail and hammer into the top of the shell. Pass the thread through the hole and knot it.

You may lightly spray your finished mobile with clear varnish or shellac to highlight its graceful beauty.

Hang it from the ceiling near a window where it can be set into motion by a breeze and reflect the sunlight.

MARINE ⚲ MOBILE

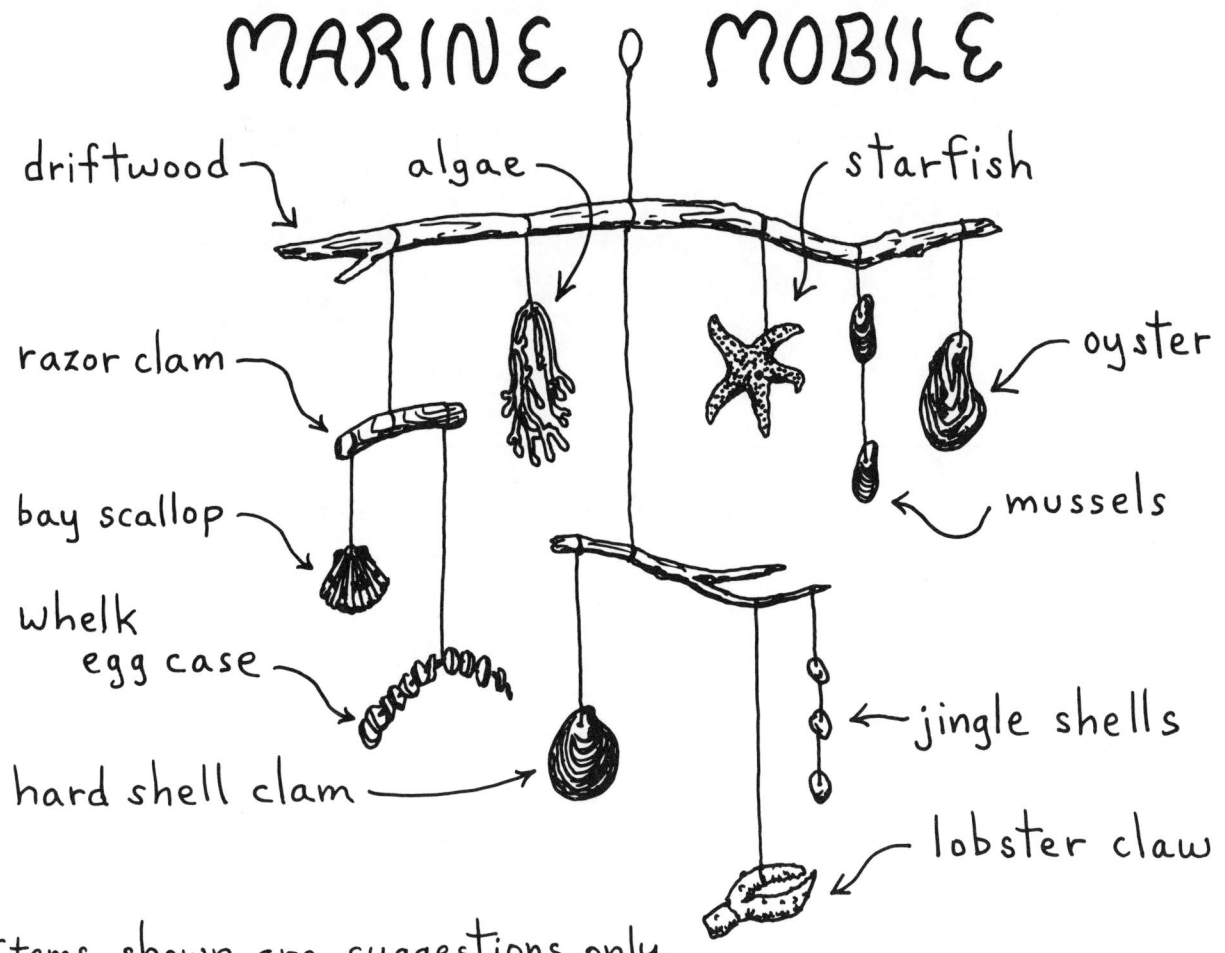

driftwood

algae

starfish

razor clam

oyster

bay scallop

mussels

whelk
egg case

hard shell clam

jingle shells

lobster claw

- Items shown are suggestions only.
- Suspend items with plastic fishing line or thread.

SAND-DRIP CASTLES

Did you ever build sand castles? You would fill your little plastic pail with damp sand, pack it tightly, then turn the pail upside-down on the beach. A careful lift of the pail would reveal a castle tower of sculpted sand. You could build moats, walls and tunnels in your sand castle. Sooner or later, though, the sun would dry out the sand and it would slowly crumble apart, or the incoming tide would rush upon the walls and they would collapse.

Here's an interesting twist on sand castle building. It takes only minutes and you don't need shovels or pails. You might even learn something about the nature and behavior of wet sand, and how well it holds its shape. These unique architectural forms are called sand-drip castles.

To build these castles, you need to work with sand that is saturated with water. The sand should have the consistency of runny oatmeal. For this reason it is best to work at the water's edge, perhaps even between waves. (If so, the life of your castle will be short, indeed. Work quickly!)

Scoop up a handful of wet sand and allow it to run in streams through your fingers as shown in the illustration.

Sand-Drip Castles

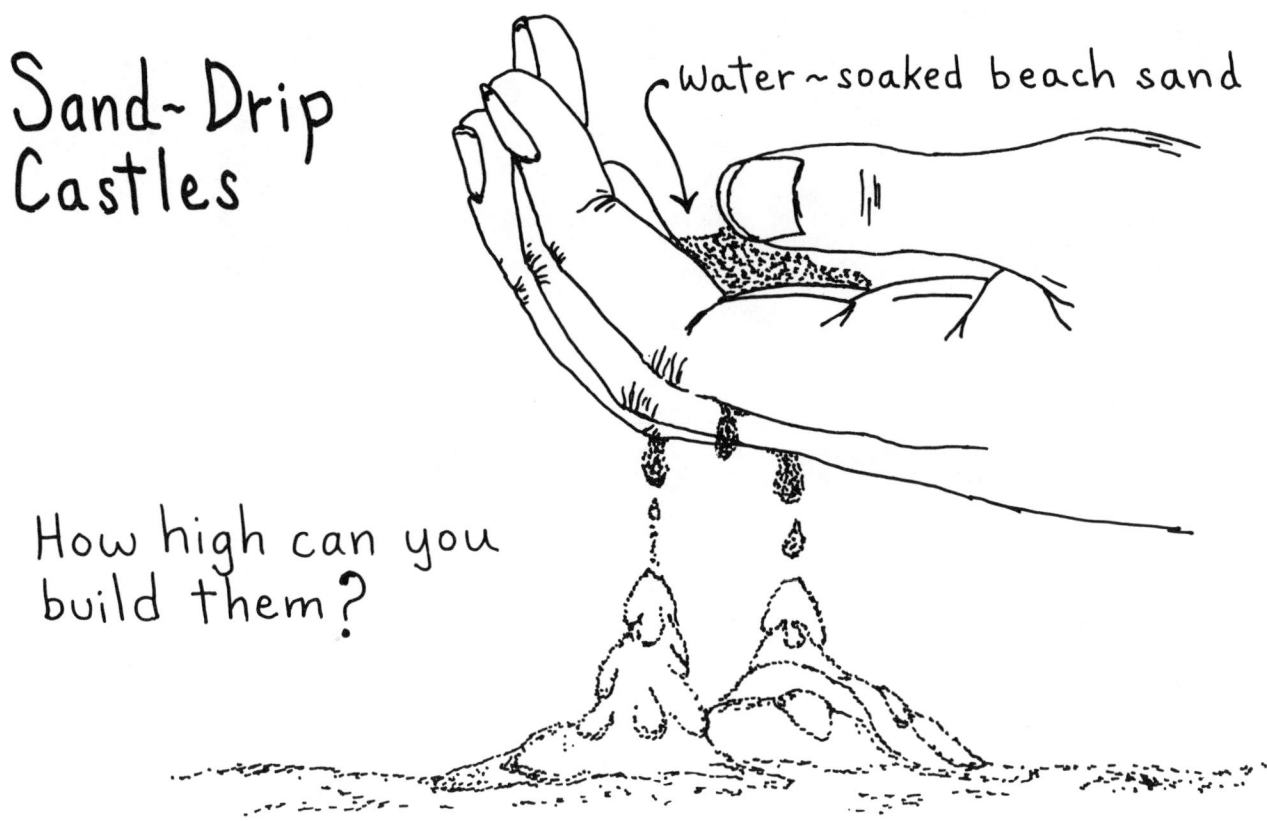

water~soaked beach sand

How high can you build them?

Use two hands at once, if you wish.

As soon as the sand has drained from your hand, scoop up more and add to your creation.

Do not shape the sand with your hands or fingers. Allow the sand to fall, spread and rise in its own natural shapes.

Try cupping your hand to let all the sand run out in a single narrow stream. How high can you get the sand to rise? Does it reach a certain height and then collapse under its own weight?

You can work further back on the beach if you fill a bucket with wet sand and add sea water until it rises above the level of the sand in the pail. This insures that your sand is saturated.

Try using funnels of different sizes, also. You can control the thickness of the sand streams and stop the flow as you wish.

Use your imagination to organize the sand drippings into architectural works of art.

MAPPING A TIDAL CREEK

Here's a wonderful little exercise that is guaranteed to perfect your skills as a cartographer, or map maker.

Many beaches and marshes contain small creeks or inlets that can be accurately mapped in a special way to produce what is called a profile map. This is a cross-section or side view of a particular area. If a salt water creek is not handy, you may have a fresh water stream in your area. This mapping activity works well in either location.

Your mapping equipment will include two sturdy stakes, perhaps four to five feet long, a ball of heavy-duty cord, a meter stick or yard stick, and an old piece of bed sheet torn into six-inch by one-inch strips. You may need a hammer or large stone for driving in your stakes. And you'll need graph paper and a pencil.

The creek you choose to map should be no more than waist-deep water, through which you can easily wade.

Do not select a creek with strong tidal currents and always work with a partner or group.

To begin, stretch your cord out on the shore and tie the rag sheet strips at accurately spaced intervals of one foot.

With this complete, tie one end of the cord to a stake. Drive this stake firmly into the sand, several feet from the water's edge, with a hammer or stone.

Tie the other end of the cord to the remaining stake and wade across the creek to the opposite shore. Drive the stake firmly into the sand several feet from the water's edge.

Wrap any excess cord around the top of the stake until it is stretched taut from shore to shore. Try to be sure that your cord is suspended level with the water's surface, if possible. This line is called your *transit line*. All your measurements will be taken with this line as your reference point.

Mapping a Tidal Creek or Inlet

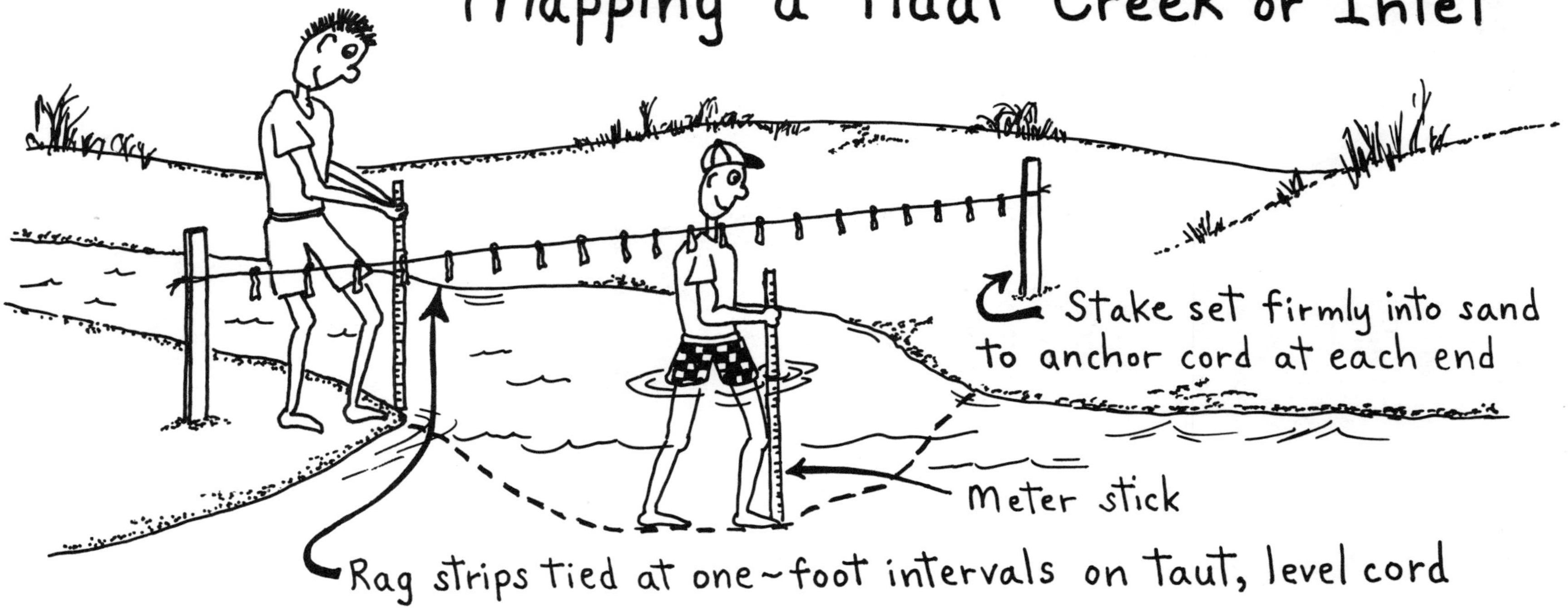

Stake set firmly into sand to anchor cord at each end
Meter stick
Rag strips tied at one-foot intervals on taut, level cord

Take vertical distances from cord to ground. Record on graph. When water's edge is reached, draw straight line across graph and take vertical readings of water's depth. Record these to map bottom profile of creek. At far shore, take distances from cord to land again.

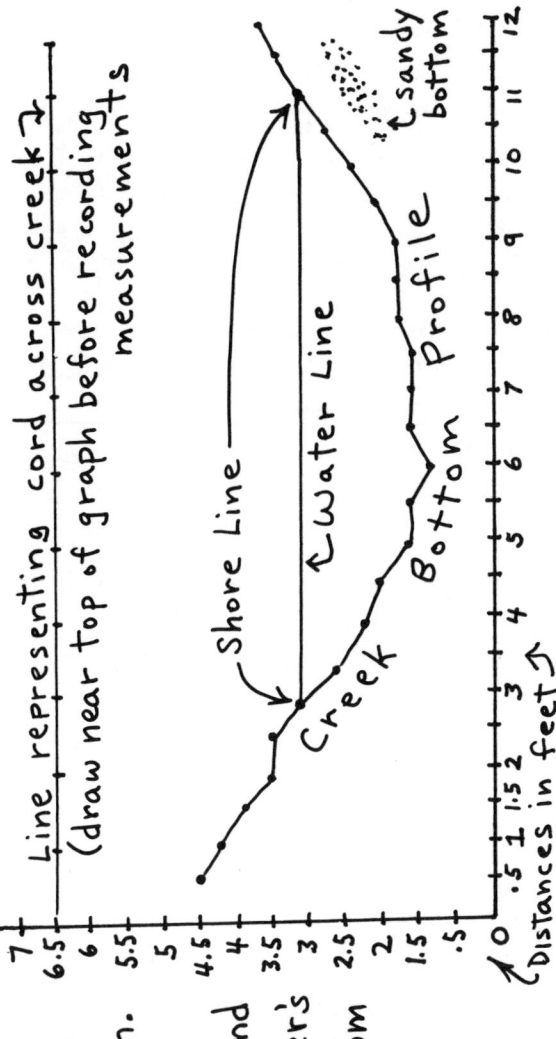

Line representing cord across creek
(draw near top of graph before recording measurements
Shore Line
Water Line
Creek Bottom Profile
sandy bottom
0
.5
1.5
2
2.5
3
3.5
4
4.5
5
5.5
6
6.5
7
.5 1 1.5 2 3 4 5 6 7 8 9 10 11 12
Distances in feet

Return to the other shore with your partner and set up your graph paper as shown in the illustration.

A clipboard will be helpful in holding your graph paper flat and providing a hard surface upon which to write.

From the edge of the first stake, begin taking vertical distances from your cord to the ground at one-foot intervals marked by the rag strips. You can be more accurate by measuring at six-inch intervals, at each rag and between each rag. As you call out the measurements, your partner can record them on the graph.

The moment your vertical measurements reach the water's edge, your task changes a bit. Draw a straight horizontal line across your graph at the point marking the water's edge. Since water level will remain constant, this will represent the surface of the water from shore to shore.

Now, take vertical measurements from the water's surface to the creek bottom, at the same intervals, and plot these points on your graph.

When you reach the opposite shore, take your measurements from the cord to the land again. At some point, your "land line" will cross the "water line" you drew.

By neatly connecting your dots on the graph, you will have an accurate profile of the creek bottom at this particular location.

You may wish to lightly color your water area blue and land area brown.

Label your map. Include the name of the location, the date, and the time of day, since tides will alter the water level. You may also include notes on the type of soil you encountered on the creek bottom, (mud, sand or stones.)

Underwater profiles can change dramatically from year to year and even season to season, due to storms, erosion and currents. Try mapping the same area sometime in the future. Is it the same?

Welcome to the club of creek cartographers!

SOLAR WATER COLLECTOR

No one plans on being marooned on a sandy island far out at sea. But it has happened to some people. What if it happened to you? Could you survive? Perhaps you could find or create shelter. And you might obtain food by catching shellfish along the shore, or by picking edible fruits from the dunes such as beach plums and rose hips. But what about fresh water? If none was readily available, how would you get it? This experiment, which uses some basic scientific principles, may give you some ideas.

There is water all around you at the beach. Unfortunately, it is sea water. You can not drink sea water. Because of its high salt content, it will actually make you thirstier because it dehydrates (removes water from) your body.

You will construct a simple device that will de-salinate, or remove the salt, from sea water to make it pure and drinkable.

To begin, you'll need a plastic cup and a black plastic garbage bag. Dark green bags will also work, though not as well as black.

At the beach, scoop out a deep, squarish pit at least six inches smaller than your garbage bag all around. Make it at least a foot deep and be sure your pit has exposed the wet beach sand below the surface.

Push the cup firmly into the sand in the center of your pit.

Spread the garbage bag over the pit opening. Be sure it extends beyond the walls of the pit on all sides.

Seal the edges of the bag by building up a continuous mound of sand all around the edges of the bag. This is to prevent the escape of moisture from the pit.

Find a small stone and place it directly in the center of the plastic. It should create a pronounced sag in the plastic directly over the opening of the cup.

Give your solar water collector an hour or so to work. Then, check your results. Carefully brush away the sand from one edge, reach in and remove the cup. Do you have sparkling fresh water? Taste it! It may be warm, but it is drinkable.

S☼LAR WATER COLLECTOR

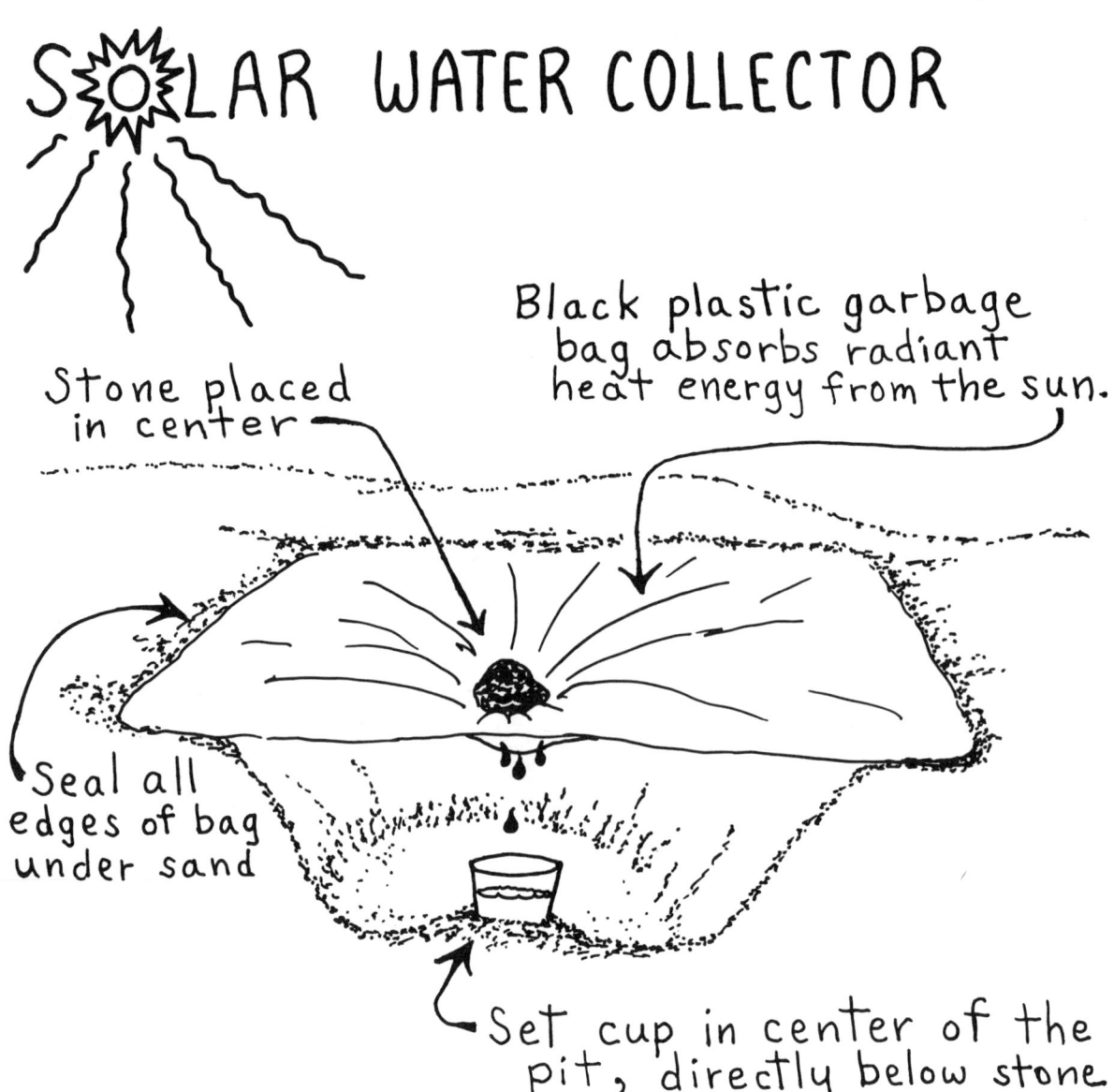

Stone placed in center

Black plastic garbage bag absorbs radiant heat energy from the sun.

Seal all edges of bag under sand

Set cup in center of the pit, directly below stone.

As heat builds up in pit, fresh water evaporates and collects (and condenses) on inside surface of plastic. Beads of water roll down the sloping surface created by the stone and drop into the cup.

How Does It Work?

The radiant energy from the sun, what we call solar energy, is absorbed by the dark color of your plastic bag. The inside of your pit heats up and causes the moisture in the sand to *evaporate*. When water evaporates, impurities like dirt and salt are left behind. As it evaporates, the water can not escape and beads of moisture collect on the inside surface of the plastic by a process called *condensation*. The tiny droplets grow larger and soon roll down the sloping plastic surface, caused by the stone you placed on top.

As the drops reach the lowest point on the plastic, they fall into the cup.

76

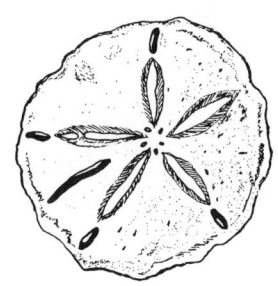

COLLECTING AND PRESERVING MARINE CREATURES

The purpose of this activity is not to have you go out and actively hunt and kill marine organisms. We can accept the fact, however, that some deaths will occur naturally in any environment. The marine environment is especially harsh. It abounds with natural predators and powerful wave forces, and is dramatically affected by winds, temperatures and other weather factors.

Often, in walking a beach, starfish, snails, shellfish, algae and other organisms are found washed ashore, already dead. These are the plants and animals you will collect. Be sure they are in a fresh state and not in the process of decomposing. If you maintain a marine aquarium with local species such as Atlantic pipe fish and sea horses, they also occasionally die and are included here.

It is a good idea to construct a simple drying rack from an old piece of screen mounted on a frame of 2" x 4" lumber stock. Air must freely circulate around the organisms being dried.

The tips that follow will be helpful in preparing and preserving many commonly found beach organisms and shells for a permanent collection...Sort of a mini-marine museum.

Algae (seaweeds) — Thin delicate algae like sea lettuce (*Ulva*), kelp (*Laminaria*), or chenille weeds (*Dasya*) may be placed in a pan of sea water. Slip an index card beneath them and lift up carefully. The algae will lay flat upon the card and adhere to the paper naturally. No glue is needed and they will retain their colors. Their shapes can be attractively arranged while still wet upon the card.

Heavier, fleshy forms of algae like sponge weed (*Codium*), rockweed (*Fucus*), or Irish moss (*Chondrus crispus*) may be air-dried on the screen for a day or two and then glued onto a mounting sheet of oaktag or onto a board. Use white glue.

Atlantic Pipefish and Sea Horses — These organisms have their skeletons on the outside, called exoskeletons, and can be air dried on the screen. Soak them in sea water to soften their bodies and gently shape them before drying. The pipefish can actually be formed into rings to fit your finger. They will be odorless and last indefinitely with care.

Small Thin-Bodied Fish — Young butterfish, sticklebacks, or similar types can be air dried on a screen in the sun. Turn them over after a few hours to thoroughly dry all sides. They will become mummified.

Starfish — These are spiny-skinned animals (*Echinoderms*). Place them in a jar of slightly diluted rubbing alcohol, about a seventy-five per cent solution, for a day or two. Remove them and boil the animal for ten minutes. Lastly, air dry them on the screen for a week or more in a shady place. The tube feet, or "suction cups," may be scraped off.

Crabs — If the shells still contain meat, place them in an ant box outdoors for a few days. The insects will pick the shells clean of meat. Then rinse the shell in alcohol and allow it to dry.

Small Snails — (mud dogs whelks, oyster drills, etc.) If any meat remains in the shell, push an old fish hook into the meat, twist firmly and remove the dead animal with a gentle pull. Wash the shell in alcohol or warm soapy water and rinse well.

Large Snails — (knobbed whelk, waved whelk, channeled whelk and conch) If animal is in shell, boil for twenty to thirty minutes. Meat will harden and shrink. Pull out entire animal with a dinner fork. (Remember that fresh whelk and conch are edible.) Wash shells in warm soapy water and rinse.

Shells of All Kinds — All empty shells, including those previously described, can be enhanced with a bright, lustrous sheen on their surface by dipping them in a solution of muriatic acid and water. Dilute the acid about fifty per cent. Muriatic acid is used by masons to clean brick and stone work of excess cement. Buy it at a mason supply or hardware store. Rinse the shell well when done

Children should not handle this material. The fumes are an irritant to eyes and nose, and the acid can cause skin irritation or burns, as well. Follow safety precautions and use the product properly.

Sand Dollars and Urchins — The skeletons of these interesting animals, called *tests*, are often washed ashore on our Atlantic beaches. They may be more common in some areas than in others. You may find one that has been bleached naturally by sun and sea, but they are commonly a pale gray to medium gray color. They can be bleached to a bright white in a fifty per cent solution of household bleach and water. Soak them for an hour and rinse well. Allow to dry. (See the related activity on using sand dollars as Christmas ornaments or medallions.)

And may your mini-marine museum be most magnificent and mystify many misters and misses for the months to come!

SAND DOLLAR CHRISTMAS ORNAMENT AND MEDALLION

The sand dollar is not only a unique marine animal, but its shell, called a *test*, makes a unique Christmas ornament or medallion.

Sand dollars belong to a group of marine animals called echinoderms, meaning "spiny skins." Other members of the group include starfish and sea urchins.

Sand dollars are rather flat, disc-shaped animals with extremely fine spines, giving their outer surface the look and feel of felt or velvet. Color may vary from brownish to purple to red. When the animal dies, the velvety coating is lost, revealing the well-known sand dollar shell, or test, with the distinctive five-petaled flower pattern on the surface. It is not a true shell. The flower pattern is created by the small pores which contained the animal's tube feet.

Sand dollars feed on tiny micro-organisms by shuffling through sandy bottoms. Their test is often found washed ashore in fall or winter, especially on northern Atlantic beaches where they tend to inhabit shallower bay waters and intertidal zones of ocean beaches.

Collect your sand dollars in the fall when fewer people walk the beach. If you can't locate any in your area, the following shops may help:

Museum Products
3175 Gold Star Highway
Old Mystic Plaza
Mystic, Connecticut 06355
(203) 536-6433

Sea Creations
Harbor Square Mall
134 Main Street
Port Jefferson, NY 11777
(516) 473-8388

Museum Products puts out a free catalog and the company offers many seashore related items for sale at reasonable prices. Call them to have it sent to you. Both shops listed above carry sand dollars.

The tests of some sand dollars found on the beach may be a dull grayish color, not yet fully bleached by the sun and sand. They may be bleached to a bright white by soaking them for one hour in a solution of water and household bleach. Use equal amounts of water and bleach. Rinse them well and and allow to dry.

Simple Christmas ornaments may be made by clipping out picture portions of used Christmas cards, perhaps a wreath, Santa's face, or a Christmas tree. Trim the picture neatly. Glue the clipped picture onto the surface of the sand dollar using clear household epoxy, sold in tubes at hardware, craft, and hobby shops.

The surface of the sand dollar is slightly rounded so hold down the edges of the picture for a moment until it sets. Allow it to dry for an hour. A clear spray may be added later.

To make a hanger, pass a six-inch length of gold elastic band or brightly colored cord through a split metal ring and tie it in a loop. Small (¼ inch) split rings are available at hardware stores and fishing tackle shops.

Glue the metal ring onto the back edge of the sand dollar with a generous drop of epoxy. Be sure it will hang with your picture right-side-up.

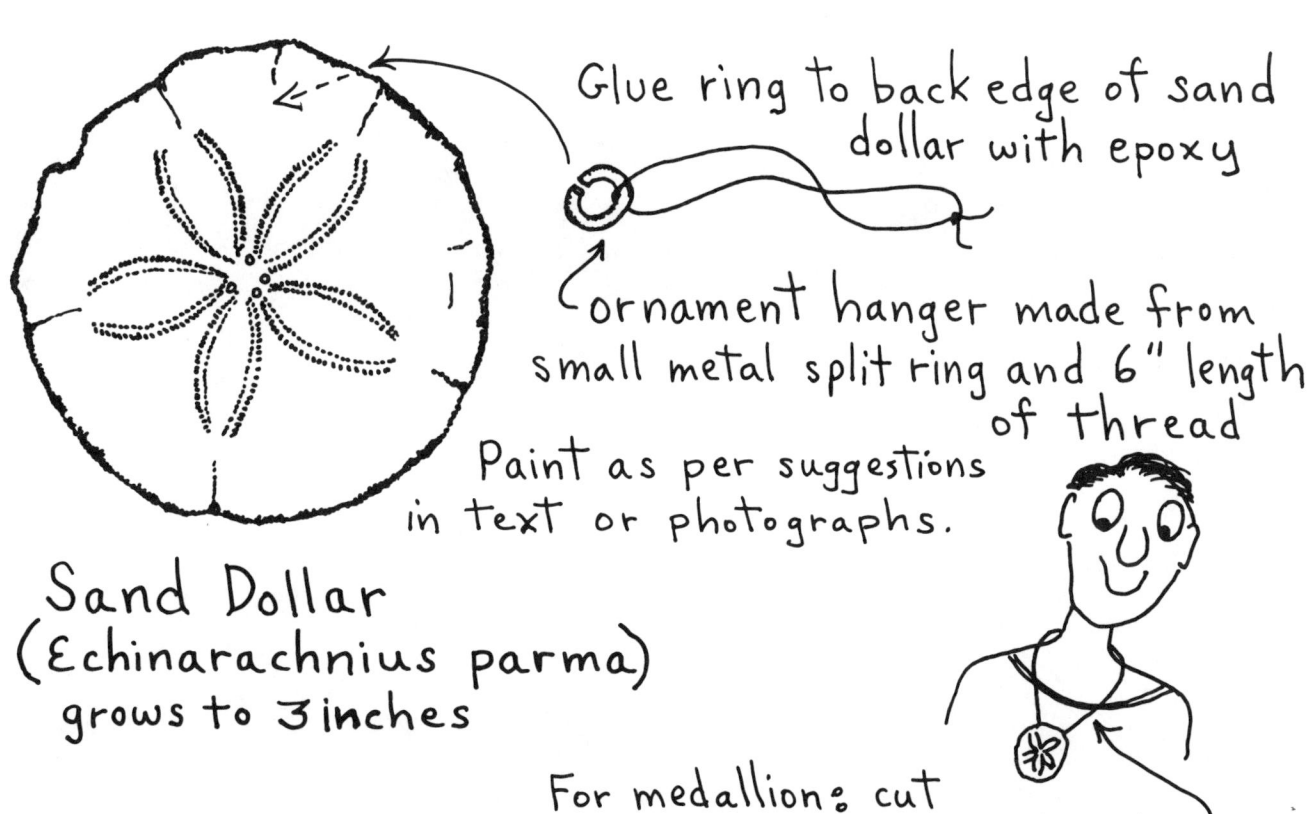

Glue ring to back edge of sand dollar with epoxy

Ornament hanger made from small metal split ring and 6" length of thread

Paint as per suggestions in text or photographs.

Sand Dollar
(Echinarachnius parma)
grows to 3 inches

For medallion: cut cord or chain to fit over head and around neck

A more creative ornament is made by hand painting. One idea, as shown in the photograph, is to paint a narrow gold band around the edge of the sand dollar. Use plastic hobby paints or acrylic enamels and a fine brush.

Next, use the five-petaled pattern, left by the sand dollar's tube feet, as guidelines and paint in the petals carefully with red paint.

Add a gold circle in the center of your "flower." Does it remind you of a Christmas poinsettia plant?

Sand dollars are quite beautiful in their natural state and can be hung as ornaments in that way. They can also be spray painted in red, green, copper, silver, and gold. Or, they may simply be varnished.

With a little thought, you may come up with creative painting ideas of your own.

For an eye-catching medallion, leave the sand dollar in its natural state (after bleaching) and attach a thin jeweler's chain or thin cord to the metal ring, long enough to fit comfortably around your neck.

"Sand Dollar Christmas Ornaments and Medallion"

82

GRASS AND LEAF PRINTS

Have you recently written a note or letter to a friend? Or maybe a piece of nature poetry?

Next time, you can personalize your writing by creating your own stationery paper with grass or leaf prints. This art activity is also a unique way of "collecting" plant specimens in a booklet of art prints.

The activity can be done at home or outdoors in the field. You'll need these materials:

1" paint brush or brayer (small rubber printing roller)
several old newspapers
9" x 12" construction or water color paper
pint of tempura paint or tube of printer's ink (any color)
tweezer, forceps, or small tongs

If you don't have all the materials available at home, they can be found at school or at any art supply shop. If you use printer's ink, you'll need a small flat board to roll out a coating of ink onto the brayer. The ink is quite thick from the tube.

Collect a variety of grasses and leaves.

Separate the newspapers into two piles, one for inking the plants and the other for making your prints, as shown in the illustration.

Place the tall grass stem or leaf to be printed in the center of the inking pile of newspaper. Brush or roll a coating of paint or ink over the entire plant surface.

Use the tweezers to lift up the grass or leaf by its stem and place it, inked side up, in the center of the printing pile of newspaper.

Take a sheet of water color paper or construction paper (white is best) and place it gently down on the inked plant. Do not let it slide or your print will smear. Press down on the paper or roll a clean brayer across the paper over the plant.

Set~up for Grass and Leaf Prints

1" brush or small rubber inking roller (brayer)

Tweezers

Pint of tempera paint (any color) or tube of printer's ink.

Ink leaves or grass on one paper pile ~ Transfer to second pile with tweezer and make print.

WATER COLOR PAPER

Gently press water color paper down onto inked leaves and lift off carefully to prevent smears.

Newspapers

Carefully lift off the paper and turn it over to examine your print. Beautiful! Let it dry.

Change the top sheet of newspaper on each pile after each print. If you are printing outdoors, be sure to discard your paper properly.

If you have chosen to make writing paper or note cards, remember to allow much free space in which to write. So, it is best to make a delicate grass print along the borders or a small leaf print at the top or bottom, perhaps in one corner. If you're producing a booklet or poster, your prints may be centered and larger.

You can even make multiple prints on a single sheet of paper in many different colors. Just be sure to let each print dry thoroughly before printing the next. You'll also have to rinse your brush or brayer between colors.

Imagine how beautiful a set of overlayed leaves in all the autumn colors would look! A print like that could be framed to decorate your home.

As a science project, make prints of the beach grasses, shrubs, or trees in your area. Identify each one using a field guide and record its name neatly on the print. Bind them together into a booklet.

Well, I'll "leaf" you alone now. You have to get started on your project.

"Grass and Leaf Prints"

SEASHORE FOOD CHAIN GAME

It is sometime difficult to understand the complex relationships that exist between organisms and the environment in which they live. We often hear warnings about disturbing the "balance of nature." This is true. Simply, if an environment is too greatly disturbed, the lives of many organisms may be threatened. In fact, it is impossible to threaten the life of one species without threatening other species due to the relationships that exist in what is called a food chain. Marine animals, in particular, are quite sensitive to changes in their environment.

The study of living things and how they interact with each other and their environment is the science of ecology. Let's look at an activity that illustrates how all forms of life are dependent upon one another for survival. It's called the food chain game.

In this activity, each player becomes a particular organism, either a plant or an animal in the marine environment, by wearing a picture card around their neck. Each card should include the full name of the plant or animal and what it eats, as well as the name of organisms that may eat, or prey upon, it. Six basic cards are provided on the following pages to serve as examples. You may make copies of these and glue them onto cardboard. Punch out the small circles and tie a twenty-four inch length of yarn to the card as a neck loop. Notice that the "barnacles" card states that these animals feed on plankton and are eaten, in turn, by blackfish.

Simple research in a seashore field guide or an encyclopedia will provide you with information to complete your own cards, if you do not already know these facts.

Add to the cards provided by drawing your own pictures or by clipping colorful close-up photographs of sea life from nature magazines.

There are a great number of sea plants and animals to choose from. Try to include some green plants, algae, in your cards, and some herbivores, or plant-eating animals. And be sure to include some carnivores, or meat-eating animals. Most carnivores are also predators, which means they hunt and kill other animals for food. The animals that are hunted and eaten are called prey.

Always include a card for humans in the food chain. We depend a great deal upon the sea for food.

You might also wish to include a card showing the sun. True, the sun is not alive, but it is solar energy that begins most food chains. Through a special process called *photosynthesis*, green plants (including algae) are able to change solar energy into chemical energy we call food.

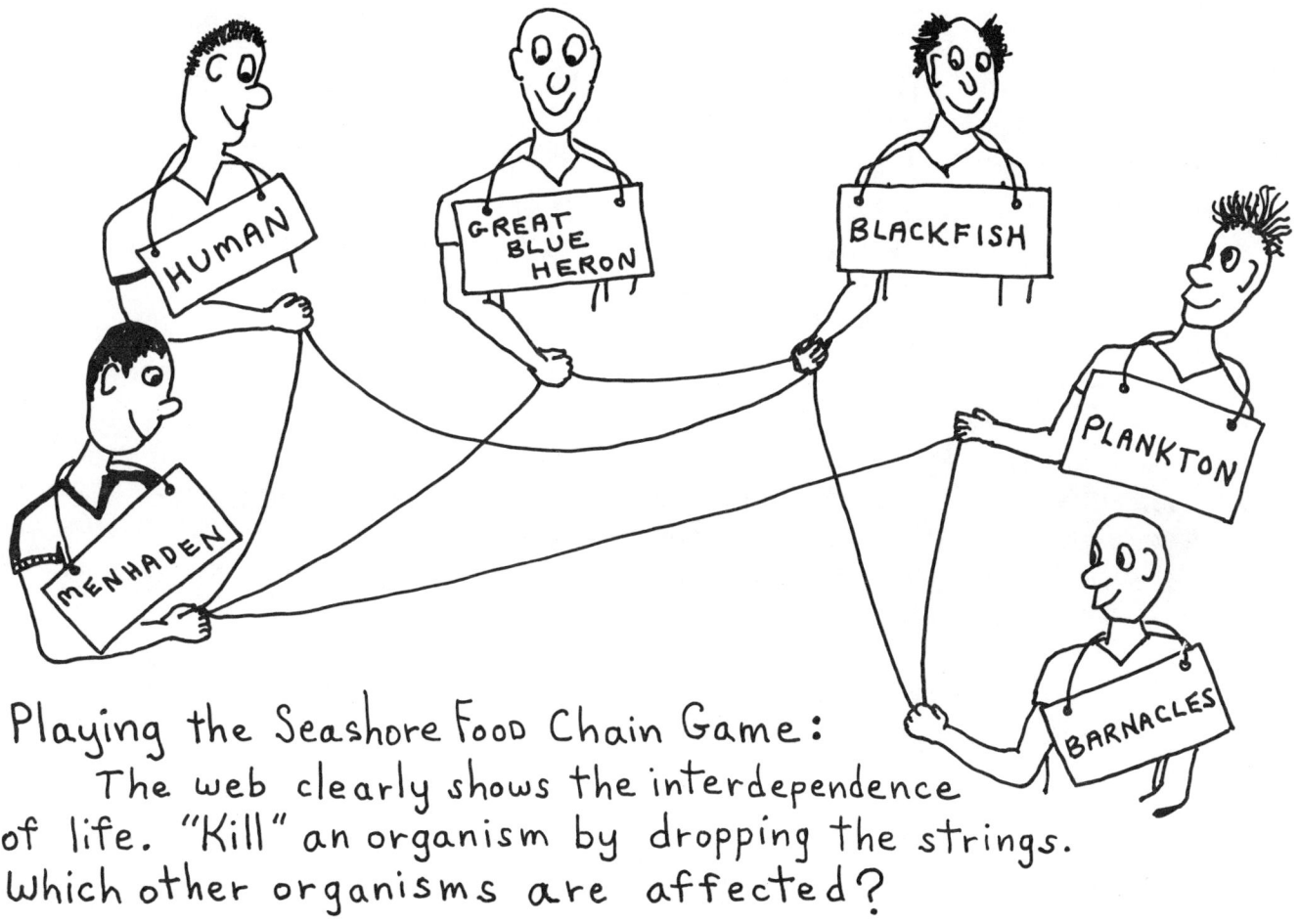

Playing the Seashore Food Chain Game:
    The web clearly shows the interdependence of life. "Kill" an organism by dropping the strings. Which other organisms are affected?

With your set of cards constructed, you're ready to play. Each person in the group selects a card from a pile that has been placed upside-down and puts the card around their neck. The players stand in a large circle, facing inward, with the picture and information on their card visible to all players.

One person has the task of cutting long lengths of yarn or heavy cord. Each end of the cord is held by any pair of organisms that are "connected" by their feeding habits. In other words, any organism that eats, or is eaten by, another organism must be connected to that organism by a length of cord.

In the example illustrated, notice that the person wearing the "barnacles" card is holding two cords. The end of one leads to the person who is the "blackfish" because these fish feed on barnacles. The other cord is held by the person wearing the "plankton" card because barnacles feed on plankton.

When all the relationships are established, you will have many food chains connected in different combinations. This is called a food web.

Keep in mind that if you have a large group of people role-playing many different organisms, your web will become quite complex. But this occurs in nature, too.

To see how these organisms are dependent upon one another for survival and how they affect each other, we will "kill" an organism by having one person drop their strings. Let's imagine that the blackfish in a certain area are suddenly exterminated by a natural or man-made disaster. There are two important effects.

First, all animals that depend upon the blackfish as a food source, like the heron, may suffer a drop in population if their food becomes scarce. They may leave the area to seek other feeding areas or they may rely on eating more of another kind of fish. If they change their diet or leave, they are being affected.

Secondly, the blackfish is a predator on barnacles. These fish actually help to control the population of barnacles. It is now possible that barnacles will overpopulate their area and cause a drop in the plankton as their own food source because there may not be enough plankton to support the large number of barnacles. And so on.

Your cards might include marine worms, snails, gulls, sharks, crabs, whales, sea lettuce, kelp, or other organisms. You can also make a set of cards to demonstrate the food chains in a dune environment.

These cards are fun and simple to make, and can help to illustrate the importance of keeping nature in balance.

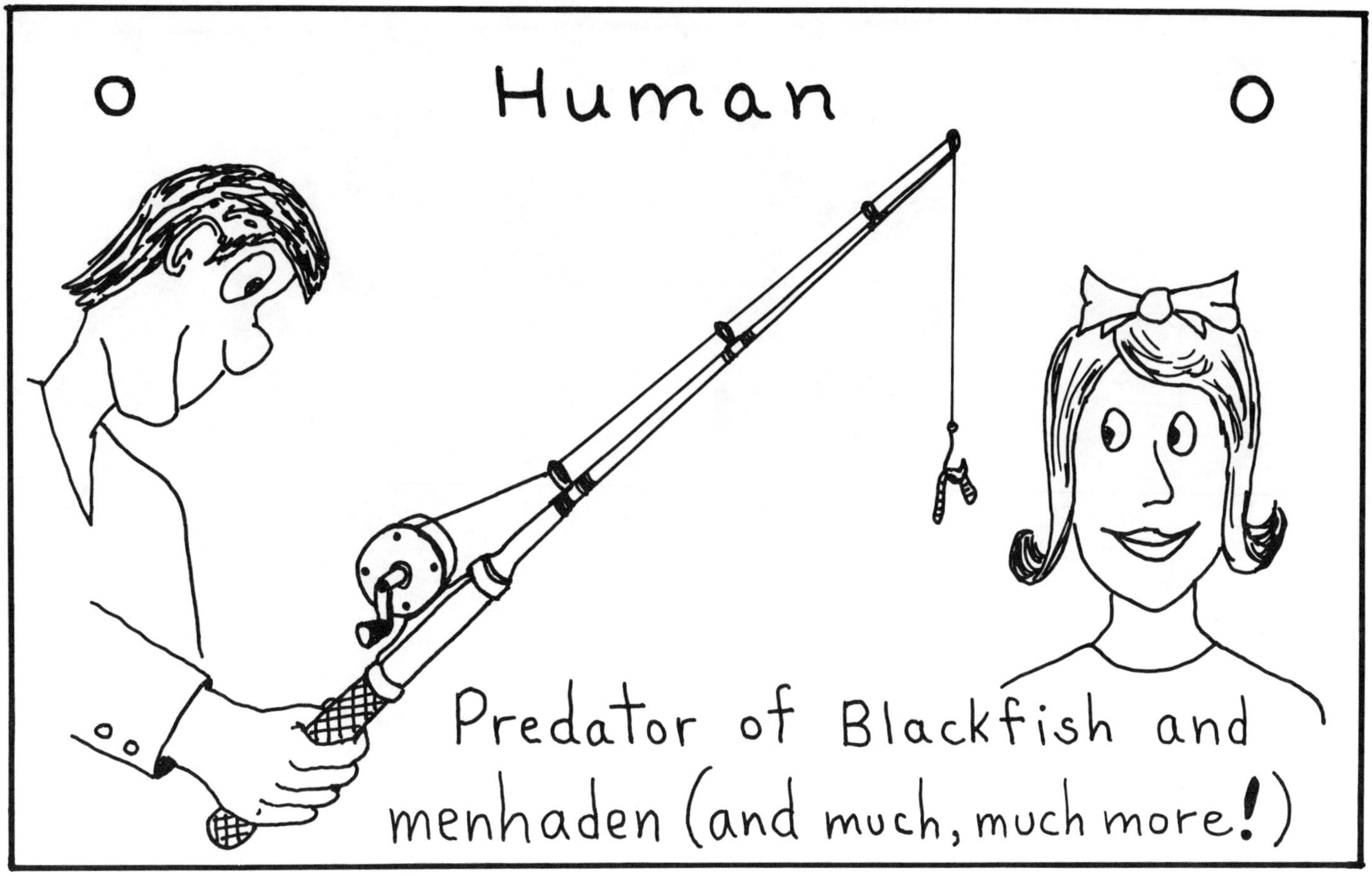

Human

O   O

Predator of Blackfish and menhaden (and much, much more!)

O   O

Great Blue Heron

Feeds on small or young fish

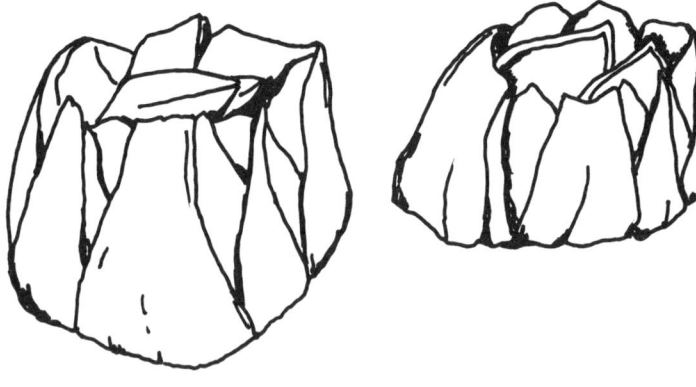

Blackfish
or Tautog

Feeds on: Barnacles/Shellfish
Eaten by: Humans

Barnacles

Feed on:
Plankton

Eaten by:
Blackfish (Tautog)

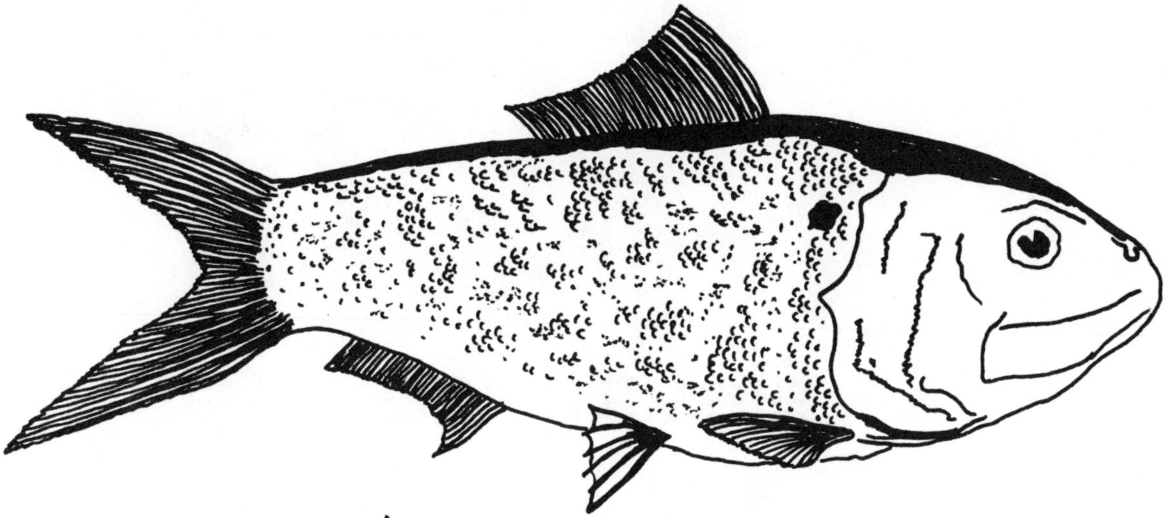

Atlantic Menhaden
(Mossbunker)

Feeds on: Plankton
Eaten by Heron and Man

Plankton
(Tiny drifting plants and animals)

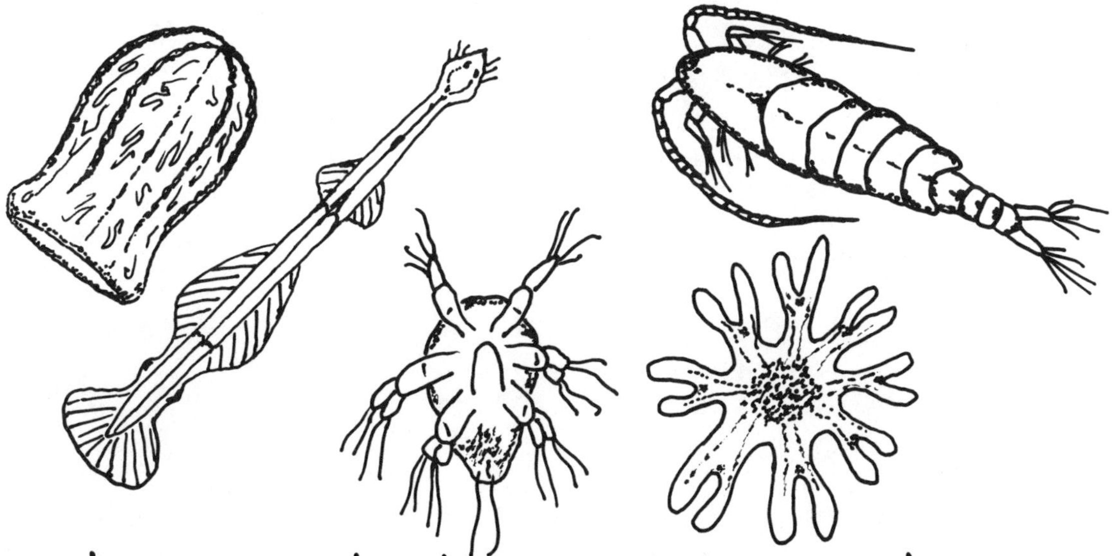

Eaten by: Menhaden and Barnacles

SHELL FLOWERS

Artists use eye and imagination in creating art. They learn to look at common objects differently, and then create the uncommon from them.

How about creating some rather uncommon flowers from the common sea shells you collect on the beach? The illustration will give you some examples.

Petals can be arranged with long curved shells like mussels and razor clams, or by overlapping smaller rounded shells like scallops.

The central part of your flower can be created from a single colorful shell, perhaps a moon snail or limpet, or with a tightly overlapped arrangement of small shells such as jingles, oyster drills, or periwinkles. The possibilities are limited only by the size of your imagination.

Clean your shells thoroughly in warm, soapy water. Rinse them well and allow them to dry.

Practice composing your shells in many ways until you are satisfied with the flower you've invented.

Glue the shells into position on pieces of white or colored oaktag, or onto wooden plaques, available at craft shops. You can even use an old board that has washed ashore, upon which you can create a complete floral arrangement or bouquet of several kinds of shell flowers.

Shell Flowers

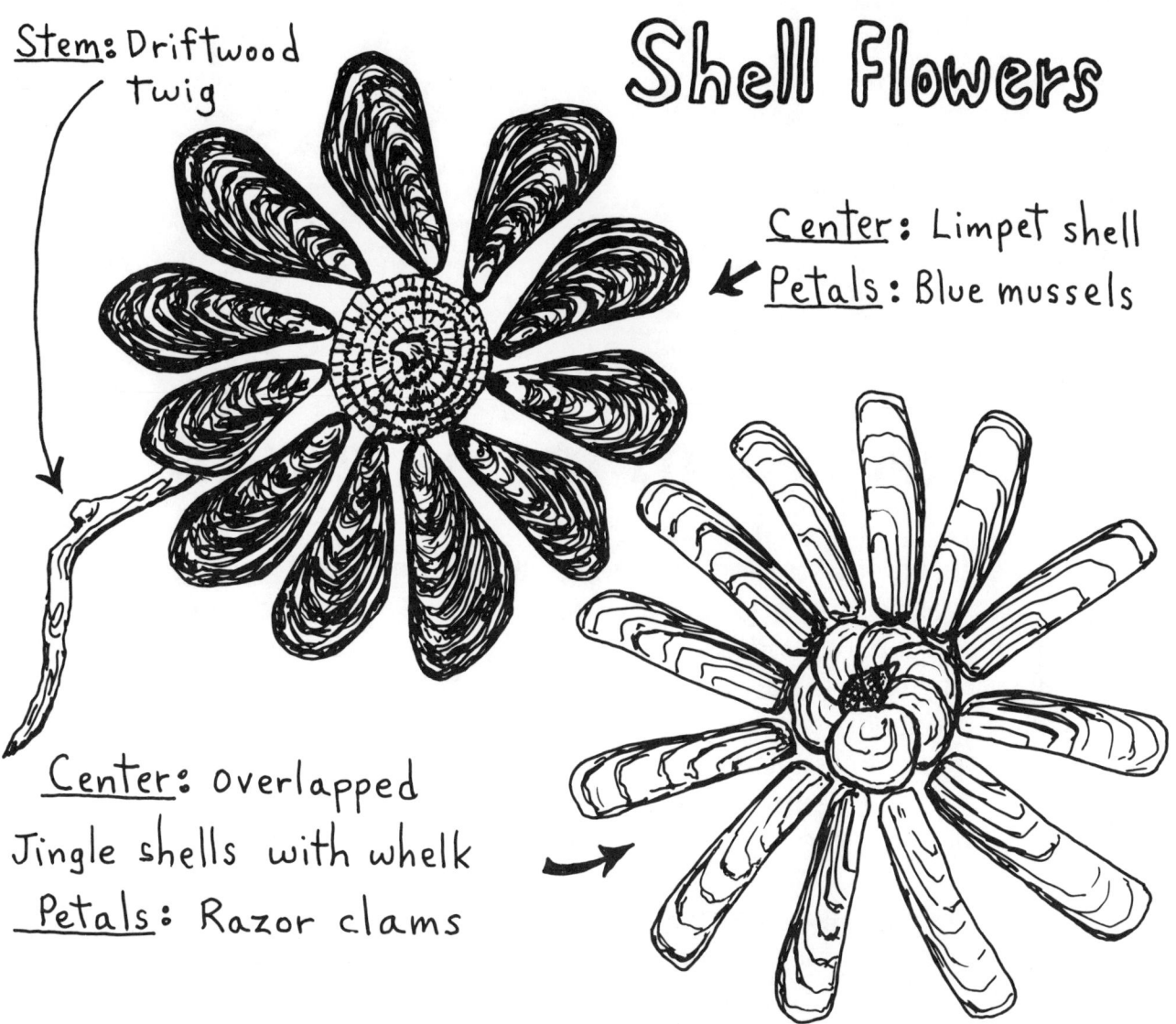

Stem: Driftwood
Twig

Center: Limpet shell
Petals: Blue mussels

Center: Overlapped
Jingle shells with whelk
Petals: Razor clams

Use white glue or a tube of household epoxy to assemble your flowers. When the glue has set, spray them with a thin coat of clear varnish or shellac to deepen the colors and add luster.

If you wish, stems can be included of driftwood twigs.

Perhaps you'd like to dream up a name for your new flower. Maybe even a fake scientific name in Latin!

Collect plenty of shells for this project. And try many arrangements and combinations of colors and shapes. Soon, an idea will blossom forth!

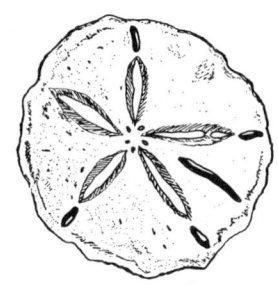

BEACH PLANTS FOR THE TASTING

As harsh an environment as the ocean beach is, our colonial ancestors found it was abundant in resources. Bayberries were boiled to remove their fragrant waxy coating and produce fine candles. Cattle were fed upon the nutritious salt-meadow hay from the marshes. Eel grass became an early form of building insulation.

The beach also provides some interesting food resources in the form of edible plants. Many plants on the beach and dunes produce fruit, usually in the form of berries, but other parts of certain plants are also edible.

All berry-producing plants are important as food resources for animals, birds in particular, but some may not be eaten by man. These include poison ivy, catbrier, and Virginia creeper. If you are unsure about a plant, *never taste the fruit.* If you wish to expand your knowledge, perhaps a school near you offers a course in field botany. Otherwise, you might try a good field guide to edible plants.

The following short list is by no means complete, but does include some of the more commonly found edible beach plants.

Sea Lettuce (*Ulva lactuca*) — is a member of the green algae family of plants. A common "seaweed," it resembles a leaf of lettuce and is often found washed ashore at the tide line. The plant is tissue-thin and its color is a bright, light green, sometimes with patchy areas of white. The plant often has irregular holes in it.

It may be rinsed and eaten raw in salads or boiled.

Sea Rocket (*Cakile edentula*) — is a member of the mustard family. Its thick fleshy leaves, stems, and rocket-shaped seed capsules have the sharp tang of horse radish or mustard. Slice it and toss it into salads for that extra zest.

Sea rocket is found growing on the upper beach and is resistant to the salt spray of the ocean. Its pale purplish flowers bloom from July through September.

Sea Lettuce

Sea Rocket

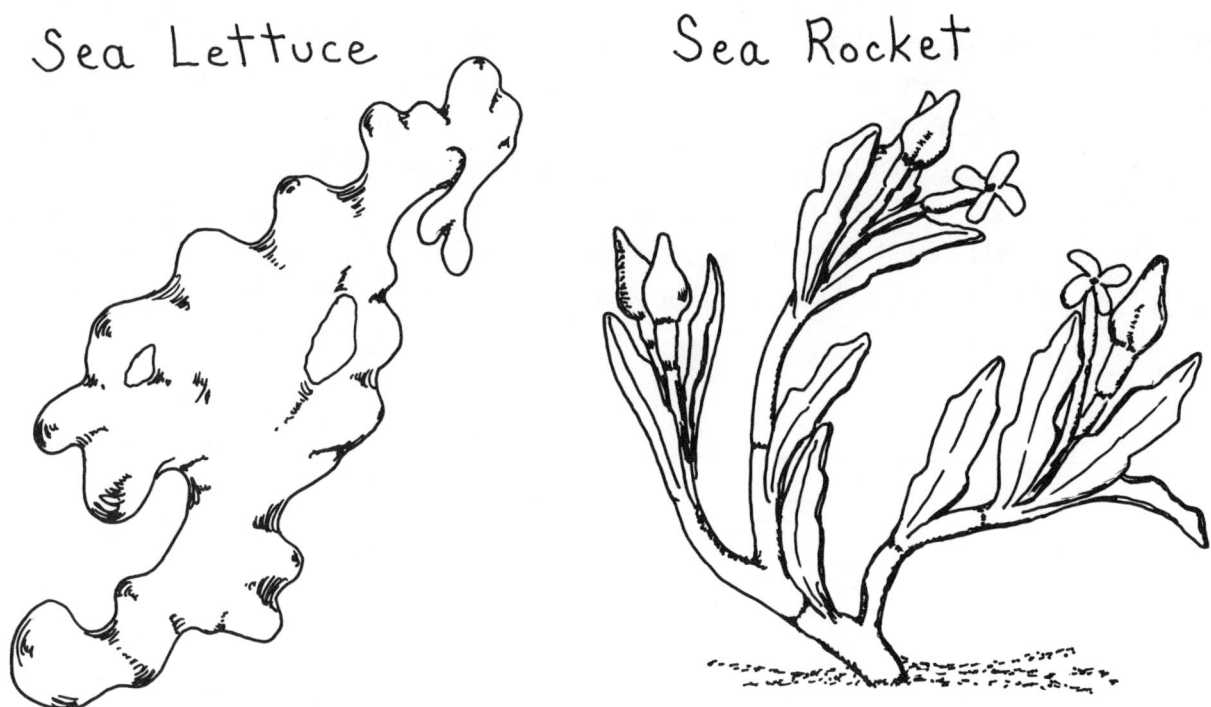

Glasswort or **Samphire** (*Salicornia europaea*) — is also known as pickle plant because colonial wives often used them in making pickles. Even today, one may come across recipes for this plant. Glasswort is a low, fleshy, attractive plant with rows of branching, sausage-like joints. Rinsed and eaten raw, it has a pleasant salty taste. Or, place the plants in a pickle brine for a few days.

Beach Pea (*Lathyrus maritimus*) — can be found growing on the crest of the primary dunes. Its small purple flowers bloom all summer, producing tiny edible pods of peas throughout the summer season. Beach peas may be eaten raw or cooked as a stir-fry vegetable.

Glasswort or Samphire

Beach Pea

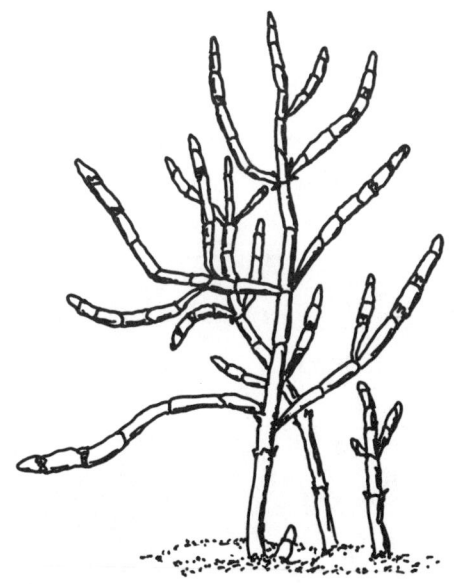

Beach Plum (*Prunus maritima*) — is a woody shrub which grows in the swale, the shallow sheltered valley behind the fore, or primary, dune. Often, they grow in low, bushy clumps on the landward side of the dune slope.

Beach plums, when fully ripened (a deep blue-black color), are a bit tart but can be eaten in their natural state. Most often they are cooked to prepare delicious beach plum jelly. Recipes for this jelly are found in many books, including *BEACHCRAFT BONANZA*, the companion guide to the book you are now reading.

Beach Plum

Other edible plants of the beach include Seabeach Orache, whose leaves may be boiled and eaten like spinach, or the rose hips of the Salt-Spray Rose, high in vitamin C and used in making rose hip jelly.

If you are interested, many written sources may further your education on beach plants. Some are listed in the bibliography. But always be sure of the plant before tasting. We don't want the fruits of your labors to be unpleasant.

SURPRISING TEMPERATURES: GRAPH ACTIVITY

"Let's go to the beach and cool off."

"Wow! It sure is hot at the beach today."

Well, what is it? Hot or cool? In fact, it's both of these and everything in between. The variety of temperatures in a beach environment is truly surprising, as this activity will clearly demonstrate.

You'll need a sturdy thermometer marked in Celsius degrees, a pencil, and the data sheet provided on the next page. You will record three temperatures in the water (ocean, bay, and tidepool), three on the beach sand at varying depths, and two in the air at different heights above the sand. You can add others such as air temperature above the ocean water. Not all categories may suit your area. Fill in only those that apply.

Follow these tips in using your thermometer to ensure accuracy and safety. Always read your thermometer at eye level and do not hold the bulb as you take your readings. Allow your thermometer to remain undisturbed in each area for two or three minutes to adjust to temperature changes accurately. If your thermometer is glass, be careful using it in the sand. Never forcefully push the thermometer into the sand. It can break. Dig a hole, placing your thermometer upright into it, and replace the sand.

When all temperatures are completed, transfer them to the graph provided. Shade in each column to the proper temperature level. You've constructed a bar graph, a sort of picture of your results. Compare the temperatures. Where was it hottest? Coolest? Do you know why?

These different temperatures are an important factor in deciding what organisms will survive in each area. Some of the small animals of the beach are sensitive to even slight differences in temperatures and moistness, especially tiny organisms living in the sand like mole crabs and sand hoppers. Others are quite hardy and can adapt, or change, as the environment changes. What animal life did you notice in each area?

SURPRISING TEMPERATURES: Data Sheet and Graph

Use your thermometer and record the temperatures on the lines provided. Some categories may not be appropriate for your area. Add your own as needed.

WATER:

Water temp. of ocean _____
Water temp. of bay _____
Water temp. of tide pool _____

BEACH:

Sand temp. one inch deep_____
Sand temp. 6 inches deep_____
Sand temp. 12 inches deep_____

AIR:

Air temp. at shoulder height _____
Air temp. one inch above sand _____

Transfer your readings above to their properly labeled columns on the graph.

Draw a line across each column at the temp. mark and shade in the columns to complete your bar graph.

Where was it hottest? Coolest? Why?

TEMPERATURE IN CELSIUS DEGREES →

42
40
38
36
34
32
30
28
26
24
22
20
18
16
14
12
10
8
6
4
2
0

WATER~ ocean
WATER~ bay
WATER~ tide pool
SAND~ one inch deep
SAND~ 6 inches deep
SAND~ 12 inches deep
AIR~ shoulder height
AIR~ 1 inch above sand

FIELD NOTES

_____

_____

_____

_____

_____

_____

_____

_____

_____

_____

_____

_____

_____

_____

_____

_____

_____

_____

_____

_____

ILLUSTRATIONS

_____

FIELD NOTES

_____
_____
_____
_____
_____
_____
_____
_____
_____
_____
_____
_____
_____
_____
_____
_____
_____
_____
_____
_____
_____
_____

ILLUSTRATIONS

_____

FIELD NOTES

_____
_____
_____
_____
_____
_____
_____
_____
_____
_____
_____
_____
_____
_____
_____
_____
_____
_____
_____
_____
_____
_____

ILLUSTRATIONS

_____

FIELD NOTES

_____
_____
_____
_____
_____
_____
_____
_____
_____
_____
_____
_____
_____
_____
_____
_____
_____
_____
_____
_____
_____
_____
_____

ILLUSTRATIONS

_____

BIBLIOGRAPHY

Amos, William H. *Life of the Seashore*.
   New York: McGraw Hill, 1966.

The Audobon Society Book of Marine Wildlife.
   New York: Harry Abrams, 1980.

Bascom, Willard. *Waves and Beaches*.
   Garden City, N.Y.: Anchor Books, Doubleday, 1980.

Berrill, N.J. *The Living Tide*.
   New York: Dodd Mead, 1953.

Berrill, N.J. and Jacquelyn Berrill. *One Thousand and One Questions Answered About the Seashore*.
   New York: Dover Publications, 1976.

Carson, Rachel. *The Edge of the Sea*.
   Boston: Houghton Mifflin Co., 1979.

Crowder, William. *Seashore Life Between the Tides*.
   New York: Dover Publications, 1975.

Dawson, E. Yale. *Marine Botany, An Introduction*.
   New York: Holt, Rinehart & Winston, 1966.

Edey, Maitland A. and the Editors of Time-Life Books. *The Northeast Coast*.
   New York: Time-Life Books, 1972.

Gosner, Kenneth L. *A Field Guide to the Atlantic Seashore*.
   Boston: Houghton Mifflin Co., 1979.

Hay, John. *The Great Beach*.
   New York: W.W. Norton & Co., 1980.

Heinz, Brian J. *Beachcraft Bonanza*.
   New York: Ballyhoo Books, 1986.

Kinsbury, J.M. *Seaweeds*.
   Massachusetts: The Chatham Press Inc., 1969.

Kopper, P. *The Wild Edge: Life and Lore of the Great Atlantic Beaches*.
   Times Books, 1979.

MacGinitie, G.E., and Nettie MacGinitie. *Natural History of Marine Animals 2nd Edition*.
   New York: McGraw Hill, 1968.

McClane, A.J. *McClane's Field Guide to Saltwater Fishes of North America*.
   New York: Holt, Rinehart & Winston, 1978.

Miner, R.W. *Field Book of Seashore Life*.
   New York: G.P. Putnam's Sons, 1950.

Petry, Loren C. and Marcia G. Norman. *A Beachcomber's Botany*.
   Chatham: The Chatham Conservation Foundation, Inc., 1968.

Sterling, Dorothy. *The Outer Lands*.
   New York: W.W. Norton & Co., Inc., 1978.

Teal, J. and M. *Life and Death of the Salt Marsh*.
   New York: Little, Brown and Company, 1969.

Ursin, Michael J. *A Guide to Fishes of the Temperate Atlantic Coast*.
   New York: E.P. Dutton, 1977.